Complete and Compact Minimal Surfaces

Mathematics and Its Applications

Volume 54

Complete and Compact Minimal Surfaces

by

Kichoon Yang

Department of Mathematics,
Arkansas State University, U.S.A.

KLUWER ACADEMIC PUBLISHERS
DORDRECHT / BOSTON / LONDON

Library of Congress Cataloging in Publication Data

Yang, Kichoon.
Complete and compact minimal surfaces / by Kichoon Yang.
p. cm. -- (Mathematics and its applications)
Includes bibliographical references.
ISBN-13:978-94-010-6947-2 e-ISBN-13:978-94-009-1015-7
DOI: 10.1007/978-94-009-1015-7

1. Surfaces, Minimal. I. Title. II. Series: Mathematics and its applications (Kluwer Academic Publishers.
QA644.Y36 1989
516.3'62--dc20 89-15576

ISBN-13:978-94-010-6947-2

Published by Kluwer Academic Publishers,
P.O. Box 17, 3300 AA Dordrecht, The Netherlands.

Kluwer Academic Publishers incorporates
the publishing programmes of
D. Reidel, Martinus Nijhoff, Dr W. Junk and MTP Press.

Sold and distributed in the U.S.A. and Canada
by Kluwer Academic Publishers,
101 Philip Drive, Norwell, MA 02061, U.S.A.

In all other countries, sold and distributed
by Kluwer Academic Publishers Group,
P.O. Box 322, 3300 AH Dordrecht, The Netherlands.

printed on acid free paper

To My Parents

SERIES EDITOR'S PREFACE

'Et moi, ..., si j'avait su comment en revenir, je n'y serais point allé.'

Jules Verne

The series is divergent; therefore we may be able to do something with it.

O. Heaviside

One service mathematics has rendered the human race. It has put common sense back where it belongs, on the topmost shelf next to the dusty canister labelled 'discarded non-sense'.

Eric T. Bell

Mathematics is a tool for thought. A highly necessary tool in a world where both feedback and non-linearities abound. Similarly, all kinds of parts of mathematics serve as tools for other parts and for other sciences.

Applying a simple rewriting rule to the quote on the right above one finds such statements as: 'One service topology has rendered mathematical physics ...'; 'One service logic has rendered computer science ...'; 'One service category theory has rendered mathematics ...'. All arguably true. And all statements obtainable this way form part of the raison d'être of this series.

This series, *Mathematics and Its Applications*, started in 1977. Now that over one hundred volumes have appeared it seems opportune to reexamine its scope. At the time I wrote

> "Growing specialization and diversification have brought a host of monographs and textbooks on increasingly specialized topics. However, the 'tree' of knowledge of mathematics and related fields does not grow only by putting forth new branches. It also happens, quite often in fact, that branches which were thought to be completely disparate are suddenly seen to be related. Further, the kind and level of sophistication of mathematics applied in various sciences has changed drastically in recent years: measure theory is used (non-trivially) in regional and theoretical economics; algebraic geometry interacts with physics; the Minkowsky lemma, coding theory and the structure of water meet one another in packing and covering theory; quantum fields, crystal defects and mathematical programming profit from homotopy theory; Lie algebras are relevant to filtering; and prediction and electrical engineering can use Stein spaces. And in addition to this there are such new emerging subdisciplines as 'experimental mathematics', 'CFD', 'completely integrable systems', 'chaos, synergetics and large-scale order', which are almost impossible to fit into the existing classification schemes. They draw upon widely different sections of mathematics."

By and large, all this still applies today. It is still true that at first sight mathematics seems rather fragmented and that to find, see, and exploit the deeper underlying interrelations more effort is needed and so are books that can help mathematicians and scientists do so. Accordingly MIA will continue to try to make such books available.

If anything, the description I gave in 1977 is now an understatement. To the examples of interaction areas one should add string theory where Riemann surfaces, algebraic geometry, modular functions, knots, quantum field theory, Kac-Moody algebras, monstrous moonshine (and more) all come together. And to the examples of things which can be usefully applied let me add the topic 'finite geometry'; a combination of words which sounds like it might not even exist, let alone be applicable. And yet it is being applied: to statistics via designs, to radar/sonar detection arrays (via finite projective planes), and to bus connections of VLSI chips (via difference sets). There seems to be no part of (so-called pure) mathematics that is not in immediate danger of being applied. And, accordingly, the applied mathematician needs to be aware of much more. Besides analysis and numerics, the traditional workhorses, he may need all kinds of combinatorics, algebra, probability, and so on.

In addition, the applied scientist needs to cope increasingly with the nonlinear world and the extra mathematical sophistication that this requires. For that is where the rewards are. Linear models are honest and a bit sad and depressing: proportional efforts and results. It is in the non-linear world that infinitesimal inputs may result in macroscopic outputs (or vice versa). To appreciate what I am hinting at: if electronics were linear we would have no fun with transistors and computers; we would have no TV; in fact you would not be reading these lines.

There is also no safety in ignoring such outlandish things as nonstandard analysis, superspace and anticommuting integration, p-adic and ultrametric space. All three have applications in both electrical engineering and physics. Once, complex numbers were equally outlandish, but they frequently proved the shortest path between 'real' results. Similarly, the first two topics named have already provided a number of 'wormhole' paths. There is no telling where all this is leading - fortunately.

Thus the original scope of the series, which for various (sound) reasons now comprises five subseries: white (Japan), yellow (China), red (USSR), blue (Eastern Europe), and green (everything else), still applies. It has been enlarged a bit to include books treating of the tools from one subdiscipline which are used in others. Thus the series still aims at books dealing with:

- a central concept which plays an important role in several different mathematical and/or scientific specialization areas;
- new applications of the results and ideas from one area of scientific endeavour into another;
- influences which the results, problems and concepts of one field of enquiry have, and have had, on the development of another.

Minimal surfaces, both with a given boundary and without boundary, are a particularly esthetically pleasing subject of mathematics. Partly this comes about because of the manifold interrelations of the subject with various parts of mathematics such as local and global differential geometry, the calculus of variations, the theory of functions, the theory of partial differential equations, topology, measure theory and algebraic geometry. On the other hand, something (natural) gets minimized and that almost immediately and inevitably means that there are interesting applications in the (physical) sciences. Lastly, the resulting geometrical shapes simply do tend to be beautiful.

It is also true that a great deal has happened in the theory of minimal surfaces in the last decennia.

The present volume gives an account of the exciting developments in recent years for the case of minimal surfaces without boundary, together with a brief look at the applications of these results to that powerful and fascinating program that goes under the name of twistor theory.

The shortest path between two truths in the real domain passes through the complex domain.

J. Hadamard

La physique ne nous donne pas seulement l'occasion de résoudre des problèmes ... elle nous fait pressentir la solution.

H. Poincaré

Never lend books, for no one ever returns them; the only books I have in my library are books that other folk have lent me.

Anatole France

The function of an expert is not to be more right than other people, but to be wrong for more sophisticated reasons.

David Butler

Bussum, July 1989

Michiel Hazewinkel

Table of Contents

Series Editor's Preface	vii
Preface	xi
Chapter I. Complete Minimal Surfaces in R^n	1
§1. Intrinsic Surface Theory	3
§2. Immersed Surfaces in Euclidean Space	8
§3. Minimal Surfaces and the Gauss Map	13
§4. Algebraic Gauss Maps	20
§5. Examples	31
§6. Minimal Immersions of Punctured Compact Riemann Surfaces	35
§7. The Bernstein–Osserman Theorem	41
Chapter II. Compact Minimal Surfaces in S^n	46
§1. Moving Frames	47
§2. Minimal Two–Spheres in S^n	52
§3. The Twistor Fibration	64
§4. Minimal Surfaces in $\mathbb{H}P^1$	71
§5. Examples	76
Chapter III. Holomorphic Curves and Minimal Surfaces in CP^n	80
§1. Hermitian Geometry and Singular Metrics on a Riemann Surface	82
§2. Holomorphic Curves in $\mathbb{C}P^n$	86
§3. Minimal Surfaces in a Kahler Manifold	95
§4. Minimal Surfaces Associated to a Holomorphic Curve	103
Chapter IV. Holomorphic Curves and Minimal Surfaces in the Quadric	110
§1. Immersed Holomorphic Curves in the Two–Quadric	110
§2. Holomorphic Curves in Q_2	119
§3. Horizontal Holomorphic Curves in SO(m)–Flag Manifolds	122
§4. Associated Minimal Surfaces	131
§5. Minimal Surfaces in the Quaternionic Projective Space	132
Chapter V. The Twistor Method	137
§1. The Hermitian Symmetric Space SO(2n)/U(m)	137
§2. The Orthogonal Twistor Bundle	140
§3. Applications: Isotropic Surfaces and Minimal Surfaces	144
§4. Self–Duality in Riemannian Four–Manifolds	149
Bibliography	153
Index	169

PREFACE

These notes contain an exposition of the theory of minimal surfaces without boundary. There have been many exciting recent developments in the study of minimal surfaces in various Riemannian manifolds and we have tried to present some of these materials from a consistent perspective. Main tools we have used are the method of moving frames and the theory of compact Riemann surfaces.

Chapter I gives a, more or less, standard treatment of minimal surfaces in $\mathbb{R}^n$ with a distinct emphasis on complete minimal surfaces with finite total curvature. In §6 we prove an immersion theorem for punctured compact Riemann surfaces resolving one of the questions raised by Osserman in his book [O6].

In Chapters II, III, and IV we concentrate on the case where the target manifold is a compact Riemannian homogeneous space (S^n, $\mathbb{C}P^n$, Q_n, and other related spaces), and the technique of complex differential geometry becomes particularly effective. The works of Calabi, Bryant, Eells–Wood, and Chern–Wolfson are presented in Chapters II and III from a uniform point of view. Chapter IV represents a generalization of these works to include minimal surfaces in complex quadrics and flag manifolds.

A cursory look at the twistor method and its applications to minimal surface theory is given in the final chapter. Of interest is our construction of 2^γ, $\gamma = n(n-1)/2$, many almost complex structures (these include the usual almost complex structures J_+ and J_-) on the orthogonal twistor bundle over a 2n–dimensional Riemannian manifold. These structures may prove to be useful in studying minimal surfaces.

Many people contributed to the completion of this writing project by sending me their preprints/reprints and by providing me with helpful comments and stimulating conversations. In particular my thanks go to N. Ejiri, H. Fujimoto, M. Guest, D. Hoffman, G. Jensen, R. Osserman, M. Rigoli, and S. Salamon. I also want to thank R. Liao who has read the entire manuscript and caught several mysterious statements. Finally it is my pleasure to thank Dr. M. Hazewinkel for writing Series Editor's Preface and for giving me several helpful suggestions; Dr. D.J. Larner and his staff for their expert handling of the manuscript.

Kichoon Yang
Arkansas State University
1989

Chapter I. Complete Minimal Surfaces in $\mathbb{R}^n$

Let M be a compact oriented smooth manifold with boundary ∂M (possibly $\partial M = \emptyset$). Also let f: $M \to (N, ds^2)$ be an immersion into a Riemannian manifold N. By a *smooth variation of f* we mean a smooth map

$$F: I \times M \to N, \; I = (-1, 1) \text{ such that}$$

i) $F_t = F(t, \cdot): M \to N$ is an immersion;

ii) $F_0 = f$;

iii) $F_t|_{\partial M} = f|_{\partial M}$ for every $t \in I$.

Fix a variation F of an immersion f and let $d\,vol_t$ denote the volume element of M with the metric induced by F_t. Define the map $A_F: I \to \mathbb{R}$ by

$$A_F(t) = \int_M d\,vol_t.$$

Definition. f: $M \to (N, ds^2)$ is called a *minimal* immersion (or minimal submanifold) if with resepect to an arbitrary variation F of f, $A_F'(0) = 0$.

A minimal immersion f is said to be *stable* if in addition to $A_F'(0) = 0$ we have $A_F''(0) > 0$ for every variation F of f. A stable minimal submanifold has the least volume amongst nearby submanifolds with the same boundary. (Historically speaking, the theory of minimal submanifolds arose in an attempt to find the surface in $\mathbb{R}^3$ of least area among those bounded by a fixed curve. This problem, called the problem of Plateau, was given a solution by Douglas and Rado in the thirties. The Plateau problem for higher dimensional submanifolds of $\mathbb{R}^n$ was given a satisfactory treatment only quite recently by Federer, Fleming, Almgren, De Giorgi, and Reifenberg. See [F] for a detailed account. Also see [G].)

Given an immersion f: $M \to (N, ds^2)$ from any smooth manifold M we let

F denote a smooth variation of $f|_{\bar{U}}$, where U is a relatively compact oriented neghiborhood in M. Call f minimal if relative to an arbitrary F of $f|_{\bar{U}}$ we have $A'_F(0) = 0$.

So called the first variational formula states that

$$A'_F(0) = c \int <H, F_* \frac{\partial}{\partial t}> d\, vol_0,$$

where H denotes the mean curvature vector of f and c is a dimensional constant. It is an immediate consequence of this formula that an immersion into a Riemannian manifold is minimal if and only if the mean curvature vector vanishes identically.

Hereafter we deal exclusively with the case where M is a smooth oriented two–manifold without boundary and $N = \mathbb{R}^n$ with the standard metric ds_E^2. Take an immersion $f\colon M \to \mathbb{R}^n$ and let Δ denote the Laplace–Beltrami operator of $(M, f^* ds_E^2)$.

Notation. $ds^2 = f^* ds_E^2$.

Local coordinates (x, y) on (M, ds^2) are called *isothermal* if

$$ds^2 = h\,(dx^2 + dy^2) \text{ for some local function } h > 0.$$

Make M into a Riemann surface by decreeing that the 1–form $dx + i dy$ is of type (1,0), where (x, y) are any isothermal coordinates. In terms of the holomorphic coordinate $z = x + iy$ we can write

$$\Delta = \frac{-4}{h}(\partial^2 / \partial z \partial \bar{z}).$$

The Gauss map of f is defined to be

$$\Phi\colon M \to \mathbb{C}P^{n-1},\ \Phi(z) = [(\frac{\partial f}{\partial z})],$$

where $[(\cdot)]$ denotes the complex line in $\mathbb{C}^n$ through the origin and $(\cdot)$. A straightforward computation shows that *f is minimal if and only if* $\Delta f = 0$, i.e., an immersion into $\mathbb{R}^n$ is minimal if and only if it is harmonic relative to the induced metric. This result coupled with the above formula for the Laplace–Beltrami operator gives

Theorem. f is minimal if and only if its Gauss map is holomorphic.

The maximum principle for harmonic functions implies that there does not exist a compact minimal surface in $\mathbb{R}^n$ and we are lead to study conformal minimal immersions from punctured compact Riemann surfaces. (Note that an immersion from an oriented two–manifold into $\mathbb{R}^n$ is, by design, conformal relative to the induced complex structure.) A conformal minimal immersion f: $M \to \mathbb{R}^n$ from a Riemann surface M is said to be *algebraic* (to be more precise, its Gauss map is algebraic) if

i) M is conformal to a compact Riemann surface M' punctured at a finite set Σ;

ii) the Gauss map Φ extends holomorphically to all of M'.

A fundamental theorem of Chern–Osserman states that *a complete conformal minimal immersion f: $M \to \mathbb{R}^n$ is algebraic if and only if the total curvature is finite.* By virtue of this theorem the theory of compact Riemann surfaces and algebraic curves is brought to bear on the study of complete minimal surfaces in $\mathbb{R}^n$ of finite total curvature.

§1. Intrinsic Surface Theory

Let M be a connected oriented smooth two–manifold. Equip M with a Riemannian metric ds^2. By a well–known theorem of Korn–Lichtenstein (see [Cher1] for a proof) we have, in a neighborhood of any point, coordinates x, y such that

$$ds^2 = h(x,y)\,(dx^2 + dy^2), \quad h > 0. \tag{1}$$

(x, y) are called *isothermal* coordinates.

Notation. $z = x + iy$.

(1) can be rewritten as

$$ds^2 = h(z)\,dz \cdot d\bar{z} \quad \text{(symmetric product)}. \tag{2}$$

Suppose we have another pair of isothermal coordinates $(\tilde{x}, \tilde{y})$ in the orientation class of (x, y). Writing $\tilde{z} = \tilde{x} + i\tilde{y}$ we have $ds^2 = \tilde{h}\,d\tilde{z} \cdot d\bar{\tilde{z}}$. Since ds^2 is globally defined we must have $h\,dz \cdot d\bar{z} = \tilde{h}\,d\tilde{z} \cdot d\bar{\tilde{z}}$. From this we compute that (Ex.)

$$d\tilde{z} = \lambda\,dz, \tag{3}$$

where λ is a $\mathbb{C}^*$–valued smooth local function on M. So z and $\tilde{z}$ are conformally related. Therefore on M there is a naturally defined complex structure, namely the complex structure whose local holomorphic coordinates are given by $x + iy$ with (x, y) isothermal and positively oriented. Hereafter given a Riemannian two–manifold (M, ds^2) we will think of it also as a Riemann surface using the aforementioned complex structure. The associated hermitian metric on M is simply

$$h\,dz \otimes d\bar{z}.$$

This time we let M be an (abstract) Riemann surface. Pick a Riemannian metric ds^2 on M and locally orthogonalize it.

$$ds^2 = (\varphi^1)^2 + (\varphi^2)^2. \tag{4}$$

$(\varphi^1, \varphi^2) = (\varphi^i)$ form a local orthonormal coframe on M. ds^2 is *compatible* with the complex structure of M if $\varphi^1 + i\varphi^2$ is a type (1,0) form on M. It is straightforward to see that any metric conformally equivalent to ds^2 (i.e., the two metrics are smooth multiples of each other) is again compatible with the complex structure and conversely any metric compatible with the complex structure of M is in fact conformally equivalent to ds^2. Thus on a Riemann surface M there exists a naturally defined conformal class of Riemannian metrics. Note that given a Riemannian two–manifold its metric is in the conformal class defined by the canonical complex structure.

Let (M, ds^2) be a Riemannian two–manifold. Given a local orthonormal frame $e = (e_i) = (e_1, e_2)$ in M define 1–forms (ω_j^i) by

(5) $$\nabla e_i = e_j \otimes \omega_i^j,$$

where ∇ denotes the Levi–Civita connection.

Convention. Hereafter by a frame (coframe) we mean an orthonormal frame (coframe) unless it is clear otherwise.

Digression (Fundamental Theorem of Riemannian Geometry). Let N be a smooth n–manifold. A connection on N is a $\mathbb{R}$–linear operator

$$\nabla: \Gamma(TN) \to \Gamma(TN \otimes T^*N)$$

satisfying the Leibnizian rule. That is to say,

$$\nabla(\xi_1 + \xi_2) = \nabla\xi_1 + \nabla\xi_2,$$
$$\nabla(f\xi) = df\cdot\xi + f\,\nabla\xi, \quad f \in C^\infty(N).$$

Suppose N is a Riemannian manifold with the metric ds_N^2. Then a Levi–Civita connection on (N, ds_N^2) is a connection which is symmetric and metric compatible, i.e.,

$$[\xi_1, \xi_2] = \nabla\xi_2(\xi_1) - \nabla\xi_1(\xi_2),$$
$$d <\xi_1, \xi_2> = <\nabla\xi_1, \xi_2> + <\xi_1, \nabla\xi_2>,$$

where [ ] is the bracket operator and < >, the inner product. The Fundamental Theorem of Riemannian Geometry states that on a Riemannian manifold there exists a unique Levi–Civita connection.

Coming back to our discussion on surfaces we let (φ^i) denote the coframe dual to (e_i). So $\varphi^i(e_j) = \delta^i_j$.

Dualizing (5) we obtain

$$d\varphi^i = -\omega^i_j \wedge \varphi^j, \tag{6}$$

where d is the exterior derivative operator.

Lemma. (ω^i_j) is $\mathfrak{o}(2)$–valued, i.e., $\omega^1_1 = \omega^2_2 = 0$, $\omega^1_2 = -\omega^2_1$.

Proof. $\langle e_i, e_j\rangle = \delta_{ij}$. Differentiate both sides and obtain

$$d\langle e_i, e_j\rangle = \langle \nabla e_i, e_j\rangle + \langle e_i, \nabla e_j\rangle = 0.$$

The result now follows from (5). □

$\omega = \omega^1_2$ is called the Levi–Civita connection form.

Notation. $\varphi = \varphi^1 + i\varphi^2$, a type (1,0) form.

Any oriented frame $\tilde{e} = (\tilde{e}_i)$ is related to $e = (e_i)$, on their common *domain, by*

$$(\tilde{e}_1, \tilde{e}_2) = (e_1, e_2)\begin{pmatrix}\cos\theta, -\sin\theta\\ \sin\theta, \cos\theta\end{pmatrix}, \tag{7}$$

where θ is a local function on M. Using the complex notation we can also write

$$\tilde{e}_1 + i\tilde{e}_2 = e^{-i\theta}(e_1 + ie_2). \tag{8}$$

It follows that $\varphi = e^{i\theta}\,\tilde{\varphi}$ and the expression $\varphi\wedge\bar{\varphi}$ is globally defined.

The Gaussian curvature of (M, ds^2), denoted by K, is given by the equation

$$d\omega = K\,\varphi^1\wedge\varphi^2 = \frac{i}{2}K\,\varphi\wedge\bar{\varphi}. \tag{9}$$

Thus K is a globally defined function on M.

We now let (x, y) be isothermal local coordinates on M. Recall that this means $z = x + iy$ is a holomorphic coordinate on M relative to the induced complex structure. Write as before $ds^2 = h\, dz \cdot d\bar{z}$. We take

(10)
$$\varphi^1 = \sqrt{h}\, dx, \quad \varphi^2 = \sqrt{h}\, dy,$$
$$e_1 = (1/\sqrt{h})\, \partial/\partial x, \quad e_2 = (1/\sqrt{h})\, \partial/\partial y.$$

Notation. $\partial/\partial z = \frac{1}{2}\,(\partial/\partial x - i\, \partial/\partial y), \quad \partial/\partial \bar{z} = \frac{1}{2}\,(\partial/\partial x + i\, \partial/\partial y).$

So $(\partial/\partial z, \partial/\partial \bar{z})$ are dual to $(dz, d\bar{z})$.

We compute that

(11)
$$\omega = \omega^1_2 = \frac{1}{2h}\,(h_y dx - h_x dy), \text{ where } h_x = \partial h/\partial x \text{ etc.}$$

The Hodge star operator on M is a linear operator

$$*\colon C^\infty(M) \to \Lambda^2(M), \quad \Lambda^2(M) \to C^\infty(M), \quad \Lambda^1(M) \to \Lambda^1(M)$$

determined by the prescription

$$1 \mapsto \varphi^1 \wedge \varphi^2 = h\, dx \wedge dy,$$
$$h\, dx \wedge dy \mapsto 1,$$
$$dx \mapsto dy, \quad dy \mapsto -dx.$$

In the above $\Lambda^i(M)$ denotes the space of exterior i–forms on M.

Using the star operator we can rewrite (11) as

(12)
$$\omega = \frac{-1}{2}\, {*d\log h}.$$

The Laplace–Beltrami operator of (M, ds^2) on $C^\infty(M)$ is defined to be

(13)
$$\Delta = -\, {*d*d}.$$

From (12) we obtain $-2d\omega = d{*d\log h}$ and

$$2{*d\omega} = -\, {*d*d\log h} = \Delta \log h.$$

But ${*d\omega} = {*(K\, \varphi^1 \wedge \varphi^2)} = K$. This gives

(14)
$$2K = \Delta \log h.$$

Let $g \in C^\infty(M)$. The verification of the following formula is routine and is left to the reader.

(15) $$\Delta g = \frac{-1}{h}(\partial^2 g/\partial x^2 + \partial^2 g/\partial y^2) = \frac{-4}{h}(\partial^2 g/\partial\bar{z}\partial z),$$

or we write

$$\Delta = \frac{-1}{h}(\partial^2/\partial x^2 + \partial^2/\partial y^2) = \frac{-4}{h}(\partial^2/\partial\bar{z}\partial z) = \frac{-4}{h}(\partial^2/\partial z\partial\bar{z}).$$

We can also express Δ in terms of an orthonormal frame (e_i).

Lemma. For $g \in C^\infty(M)$ we have

(16) $$\Delta g = \sum_i [(\nabla e_i(e_i))g - e_i(e_i g)].$$

Proof. As in (10) put $e_1 = (1/\sqrt{h})\ \partial/\partial x$ and $e_2 = 1/\sqrt{h}\ \partial/\partial y$, where $ds^2 = h\ (dx^2 + dy^2)$. Have

$$(\nabla e_i(e_i))g = [e_j \otimes \omega_i^j(e_i)]g.$$

So,

$$\sum_i (\nabla e_i(e_i))g = (-1/2h^2)h_y\ \partial g/\partial y - (1/2h^2)h_x\ \partial g/\partial x \quad \text{using (11).}$$

On the other hand,

$$\sum_i e_i(e_i g) = (1/\sqrt{h})\ \partial[(1/\sqrt{h})(\partial g/\partial x)]/\partial x + (1/\sqrt{h})\ \partial/[(1/\sqrt{h})(\partial g/\partial y)]/\partial y$$

$$= \frac{1}{h}(\partial^2 g/\partial x^2 + \partial^2 g/\partial y^2) + (1/\sqrt{h})\ \partial g/\partial x\ \partial(1/\sqrt{h})/\partial x + (1/\sqrt{h})\ \partial g/\partial y\ \partial(1/\sqrt{h})/\partial y.$$

(15) does the rest. □

§2. Immersed Surfaces in Euclidean Space

Consider a smooth immersion $f\colon M \to \mathbb{R}^n$, where M is a two–manifold. The derivative map of f is injective everywhere and by the Inverse Function Theorem f is locally an embedding. We make M into a Riemannian manifold by endowing it with the induced metric $f^* ds_E^2$, where ds_E^2 denotes the Euclidean metric.

Notation. $f^* ds_E^2 = ds^2$.

Thinking of (M, ds^2) as a Riemann surface the map becomes a conformal immersion. (This just means that the induced metric on M is in the conformal class of M which is true by construction.)

Let $U \subset M$ be a neighborhood such that $f|_U$ is an embedding. We use the phrase "locally" to mean "working inside U", and so forth.

Let $(\epsilon_i) = (\epsilon_1, \epsilon_2)$ be a (local) frame in M. So for some isothermal coordinates (x, y) we have

$$(1) \qquad \epsilon_1 = (1/\sqrt{h})\ \partial/\partial x, \quad \epsilon_2 = (1/\sqrt{h})\ \partial/\partial y,$$

where $ds^2 = h(dx^2 + dy^2)$.

Index Convention. $1 \le i,j,k,\cdots \le 2, \quad 3 \le a,b,c,\cdots \le n, \quad 1 \le \alpha,\beta,\gamma,\cdots \le n.$

A local adapted frame (along f) is a frame $(e_i;e_a)$ defined on f(U) such that (e_i) are tangential to f(U). It follows that for some isothermal coordinates x, y

$$(2) \qquad e_i = f_* \epsilon_i.$$

From now on we fix (x, y) so that (2) holds.

Notation. $p \in U$, $f(p) = q \in \mathbb{R}^n$.

Let $e = (e_\alpha) = (e_i;e_a)$ be an adapted frame in f(U). A technically important observation is that given any $q \in f(U)$ there exists a neighborhood $\tilde{U}$ of q in $\mathbb{R}^n$ and a frame $(\tilde{e}_\alpha)$ in $\tilde{U}$ such that

$$(\tilde{e}_\alpha)|_{\tilde{U} \cap f(U)} = (e_\alpha).$$

∇_E denotes the Euclidean connection on $\mathbb{R}^n$.

$$\nabla_E : \Gamma(T\mathbb{R}^n) \to \Gamma(T\mathbb{R}^n \otimes T^*\mathbb{R}^n),$$

$$(3) \qquad \nabla_E \tilde{e}_\alpha = \tilde{e}_\beta \otimes \Omega^\beta_\alpha,$$

where $\tilde{e}$ is a local frame in $\mathbb{R}^n$.

Write $\tilde{e}_\alpha = \tilde{e}^\beta_\alpha E_\beta$, where $(E_\beta) = E$ is the Euclidean frame. Then

$$\nabla_E \tilde{e}_\alpha = \nabla_E(\tilde{e}^\beta_\alpha E_\beta) = E_\beta \otimes d\tilde{e}^\beta_\alpha + 0 = \tilde{e}_\beta \otimes (\tilde{e}^{-1})^\beta_\gamma \, d\tilde{e}^\gamma_\alpha.$$

Thus we can write

(4) $$(\Omega^\alpha_\beta) = \Omega = \tilde{e}^{-1} d\tilde{e}.$$

Ω is $\mathfrak{o}(n)$–valued since ∇_E is a metric connection.

Let $e = (e_i; e_a)$ be an adapted frame along f and also let ∇ denote the Levi–Civita connection of (M, ds^2). We have

(5) $$\nabla \epsilon_i = \epsilon_j \otimes \omega^j_i,$$

(6) $$\nabla_E e_i = e_j \otimes \Omega^j_i + e_a \otimes \Omega^a_i,$$

where in (6) we use extensions $(\tilde{e}_\alpha)$ to make the computation. (The reader should verify that the computation is independent of the choice of extensions.)

Since f is an isometric immersion we must have

(7) $$f^* \Omega^j_i = \omega^j_i,$$

i.e., the tangential component of ∇_E pulls back to ∇. We also have

$$\nabla_E e_a = e_i \otimes \Omega^i_a + e_b \otimes \Omega^b_a.$$

$e_i \otimes \Omega^i_a$ (the tangential component of $\nabla_E e_a$) is usually denoted by $-A(e_a)$ and the equations

(8) $$\nabla_E e_a + A(e_a) = e_b \otimes \Omega^b_a$$

are called the Weingarten equations.

Let $(\tilde{\varphi}^\alpha)$ denote the local coframe in $\mathbb{R}^n$ dual to $(\tilde{e}_\alpha)$.

Notation. $f^* \Omega^\alpha_\beta = \omega^\alpha_\beta$, $f^* \tilde{\varphi}^\alpha = \varphi^\alpha$.

Since the e_a's are perpendicular to f(U) we see that

(9) $$\varphi^a = 0.$$

Using the fact that $\tilde{\varphi}^\alpha(\tilde{e}_\beta) = \delta^\alpha_\beta$ and (4) we see that

(10) $$d\tilde{\varphi}^\alpha = -\Omega^\alpha_\beta \wedge \tilde{\varphi}^\beta.$$

Have $d\varphi^\alpha = d \overset{*}{f} \tilde{\varphi}^\alpha = \overset{*}{f} d\tilde{\varphi}^\alpha$. So $0 = d\varphi^a = -\omega^a_i \wedge \varphi^i$ and it follows that (by Cartan's Lemma if one likes)

(11) $$\omega^a_i = S^a_{ij}\varphi^j,$$

for some local functions $S^a_{ij} = S^a_{ji}$ on M.

Put $II^a_{ij} = \Omega^a_j(e_i)$. $e_a \otimes II^a_{ij}$ is the normal component of $(\nabla_E e_j)(e_i)$. From (11) we see that

(12) $$\overset{*}{f} II^a_{ij} = II^a_{ij} \circ f = S^a_{ij} = S^a_{ji} = \overset{*}{f} II^a_{ji}.$$

The *a−th second fundamental form of f* is defined to be

(13) $$II^a = II^a_{ij}\ \tilde{\varphi}^i \cdot \tilde{\varphi}^j \text{ (symmetric product)},$$

and the *second fundamental tensor of f* is defined to be

(14) $$II = e_a \otimes II^a.$$

The *a−th mean curvature of f* is

(15) $$H^a = \tfrac{1}{2}\,(II^a_{11} + II^a_{22}) = \tfrac{1}{2} \operatorname{tr} II^a.$$

The *mean curvature vector of f* is

(16) $$H = e_a \otimes H^a.$$

Put

(17) $$\Theta^a_b = -\Omega^a_i \wedge \Omega^i_b = d\Omega^a_b + \Omega^a_c \wedge \Omega^c_b.$$

(Θ^a_b) are the *normal curvature forms of f.* (Keep in mind that these forms are defined relative to an adapted frame along f.)

Notation. $\overset{*}{f}\Theta^a_b = \theta^a_b$.

Define local functions R^a_{bij} on M by

(18) $$\theta^a_b = \tfrac{1}{2} R^a_{bij}\ \varphi^i \wedge \varphi^j \text{ (R is antisymmetric in i, j.)}$$

(11) now gives

(19) $$R^a_{bij} = S^a_{ki}S^b_{kj} - S^a_{kj}S^b_{ki}.$$

The above equations are called the Ricci equations.

Lemma. Let K denote the Gaussian curvature of (M, ds^2). Then

(20) $$K = \sum_a \det S^a, \quad S^a = (S^a_{ij}).$$

Proof. Have $\omega = \omega^1_2 = f^*\Omega^1_2$ and $d\omega = K\, \varphi^1 \wedge \varphi^2$. But

$$d\omega^1_2 = \omega^1_a \wedge \omega^a_2 = \sum_a \omega^1_a \wedge \omega^a_2 = \sum_a S^a_{1i}S^a_{2j}\, \varphi^i \wedge \varphi^j = \sum_a \det S^a\, \varphi^1 \wedge \varphi^2. \quad \square$$

Proposition. Let $f\colon (M, ds^2) \to \mathbb{R}^n$ be an isometric immersion and also let $\operatorname{tr} S^a$ denote the trace of (S^a_{ij}). ($\operatorname{tr} S^a$ is a globally defined function on M and $2 \cdot f^* H^a = \operatorname{tr} S^a$.) Then

$$-e_a(q) \otimes (\operatorname{tr} S^a)(p) = E_\alpha(q) \otimes \Delta f^\alpha(p),$$

where $f(p) = q$, $(E_\alpha = \partial/\partial x^\alpha)$ is the Euclidean frame and $f = (f^\alpha) = E_\alpha \otimes f^\alpha$. *Identifying* p with q (and $U \subset M$ with $f(U) \subset \mathbb{R}^n$) we can write

$$-2e_a \otimes H^a = (\Delta f^\alpha),$$

or simply

(21) $$-2H = \Delta f.$$

Proof. Have $\Delta f = E_\alpha \otimes \frac{-1}{h} (\partial^2 f^\alpha/\partial x^2 + \partial^2 f^\alpha/\partial y^2)$, and $2H = e_a \otimes (S^a_{11} + S^a_{22})$. Thus (21) can be rewritten as

(22) $$\frac{1}{h} (\partial^2 f^\alpha/\partial x^2 + \partial^2 f^\alpha/\partial y^2) = (S^a_{11} + S^a_{22})\, e^\alpha_a,$$

where $e_a = e^\alpha_a E_\alpha$ and we think of each e^α_a as a local function on M. (So by e^α_a we really mean $e^\alpha_a \circ f$.) From (4) we get

(23) $$S^a_{ij} = \omega^a_j(\epsilon_i) = (e^{-1})^a_\alpha\, de^\alpha_j(e_i).$$

We now express the e_i's in terms of the Euclidean frame and obtain

(24) $$e_1 = f_*\epsilon_1 = (1/\sqrt{h}) f_*(\partial/\partial x) = (1/\sqrt{h})(\partial f^\alpha/\partial x) E_\alpha, \text{ and likewise}$$

$$e_2 = (1/\sqrt{h})(\partial f^\alpha/\partial y)E_\alpha.$$

(22) now follows upon substituting the right hand sides of (24) into (23). □

§3. Minimal Surfaces and the Gauss Map

A conformal immersion $f: M \to \mathbb{R}^n$ is said to be *minimal* if the mean curvature vector H vanishes identically. By §2 (21) we have

(1) $$f \text{ is minimal iff } \Delta f = 0.$$

Corollary. There does not exist a conformal minimal immersion $f: M \to \mathbb{R}^n$ from a compact Riemann surface M.

Proof. Suppose $f: M \to \mathbb{R}^n$ is such a map. Then each f^α is a harmonic function. The maximal principle for harmonic functions states that a harmonic function on a compact surface (without boundary) must be a constant and this would imply that f is a constant map. □

Hereafter $f: M \to \mathbb{R}^n$ denotes a conformal minimal immersion from a Riemann surface M. In terms of a local holomorphic coordinate z on M we have $\Delta = \frac{-4}{h}\,\partial^2/\partial\bar{z}\partial z$, where $ds^2 = h\, dz\cdot d\bar{z}$ is the induced metric on M. The minimality of f says that

(2) $$\partial^2 f^\alpha/\partial\bar{z}\partial z = 0.$$

Put

(3) $$\eta^\alpha(z) = \partial f^\alpha/\partial z.$$

By (2) we then have

(4) $$\partial\eta^\alpha/\partial\bar{z} = 0,$$

i.e., η^α is a (local) holomorphic function on M.

Put

(5) $$\zeta^\alpha = \eta^\alpha(z)dz.$$

Lemma. Each ζ^α is a *globally* defined holomorphic 1–form on M.

Proof. let $\tilde{z}$ be another local holomorphic coordinate on M and write $\tilde{\zeta}^\alpha = \tilde{\eta}^\alpha(\tilde{z})d\tilde{z}$. Then

(6) $$\tilde{\eta}^\alpha = \partial f^\alpha/\partial\tilde{z} = (\partial f^\alpha/\partial z)(dz/d\tilde{z}) = \eta^\alpha(dz/d\tilde{z})$$

and the result follows. □

Therefore, given a conformal minimal immersion $f\colon M \to \mathbb{R}^n$ there arise n holomorohic 1–forms $\zeta_f = (\zeta^1, \cdots, \zeta^n)$ on M.

Proposition. Let f and ζ_f be as in the preceding paragraph. Then we have

(7) $$\Sigma\, |\eta^\alpha|^2 = \frac{h}{2}\,, \text{ where } ds^2 = h\ dz\cdot d\bar{z};$$

(8) $$\Sigma\, (\eta^\alpha)^2 = 0;$$

(9) the ζ^α's have no real periods, i.e., Re $\int \zeta^\alpha$ is path–independent.

Proof. $ds^2 = h\ dz\cdot d\bar{z} = h(dx^2 + dy^2)$. So $h(z) = \langle f_*\frac{\partial}{\partial x}, f_*\frac{\partial}{\partial x}\rangle = \langle f_*\frac{\partial}{\partial y}, f_*\frac{\partial}{\partial y}\rangle$ and $0 = \langle f_*\frac{\partial}{\partial x}, f_*\frac{\partial}{\partial y}\rangle$, where $\langle\ ,\ \rangle$ denotes the Euclidean inner product. Now $\langle f_*\frac{\partial}{\partial x}, f_*\frac{\partial}{\partial x}\rangle = \langle \frac{\partial f^\alpha}{\partial x} E_\alpha, \frac{\partial f^\beta}{\partial x} E_\beta\rangle = \sum_\alpha (\frac{\partial f^\alpha}{\partial x})^2 = \sum_\alpha (\frac{\partial f^\alpha}{\partial y})^2$. On the other hand, $\sum_\alpha |\eta^\alpha|^2 = \Sigma\, |\frac{\partial f^\alpha}{\partial z}|^2 = \Sigma\, (\partial f^\alpha/\partial z)(\partial f^\alpha/\partial\bar{z}) = \frac{1}{4}\Sigma\, [(\frac{\partial f^\alpha}{\partial x})^2 + (\frac{\partial f^\alpha}{\partial y})^2]$ and (7) follows. Now $\Sigma\, (\eta^\alpha)^2 = \Sigma\, (\frac{\partial f^\alpha}{\partial z})^2 = \frac{1}{4}\Sigma\, (\frac{\partial f^\alpha}{\partial x} - i\frac{\partial f^\alpha}{\partial y})^2 = \frac{1}{4}\Sigma\, [(\frac{\partial f^\alpha}{\partial x})^2 - (\frac{\partial f^\alpha}{\partial y})^2 - 2i\frac{\partial f^\alpha}{\partial x}\frac{\partial f^\alpha}{\partial y}]$ $= \frac{-i}{2}\Sigma\, \frac{\partial f^\alpha}{\partial x}\frac{\partial f^\alpha}{\partial y} = \frac{-i}{2}\langle f_*\frac{\partial}{\partial x}, f_*\frac{\partial}{\partial y}\rangle = 0$. This proves (8). To prove (9) consider $\int_\Gamma \zeta^\alpha$ where Γ is a path in M. Now

$$\zeta^\alpha = \frac{\partial f^\alpha}{\partial z}dz = \frac{1}{2}\left(\frac{\partial f^\alpha}{\partial x} - i\frac{\partial f^\alpha}{\partial y}\right)(dx + idy) = \frac{1}{2}df^\alpha + \frac{i}{2}\left(\frac{\partial f^\alpha}{\partial x}dy - \frac{\partial f^\alpha}{\partial y}dx\right).$$

That is,

$$\zeta^\alpha = \frac{1}{2}(df^\alpha + i\,{}^*df^\alpha). \tag{10}$$

We may assume that $f(z_0) = 0 \in \mathbb{R}^n$. Then we must have

$$2\,\mathrm{Re}\int_{z_0}^{z}\zeta^\alpha = f^\alpha(z) \tag{11}$$

and (9) follows. □

The following theorem enables us to manufacture minimal surfaces from holomorphic 1–forms.

Theorem. Let M be a Riemann surface. Suppose we have n holomorphic 1–forms (ζ^α) on M satisfying

$$\Sigma\,|\eta^\alpha|^2 > 0, \text{ where } \zeta^\alpha = \eta^\alpha dz \text{ locally;} \tag{12}$$

$$\Sigma\,(\eta^\alpha)^2 = 0; \tag{13}$$

$$(\zeta^\alpha) \text{ have no real periods.} \tag{14}$$

Then

$$f = (f^\alpha) = 2\,\mathrm{Re}\int_{z_0}^{z}(\zeta^\alpha): M \to \mathbb{R}^n \tag{15}$$

is a conformal minimal immersion with $f(z_0) = 0$.

Proof. From (15) we compute that $\frac{\partial f^\alpha}{\partial z} = \eta^\alpha$. f is an immersion iff $\Sigma\,(\frac{\partial f^\alpha}{\partial x}) + (\frac{\partial f^\alpha}{\partial y}) > 0$ iff $\Sigma\,|\frac{\partial f^\alpha}{\partial z}| > 0$ iff (12) holds, where $z = x + iy$. (13) now says that $\Sigma\,(\eta^\alpha)^2 = \Sigma\,(\frac{\partial f^\alpha}{\partial z})^2 = 0$. Since $\frac{\partial}{\partial z} = \frac{1}{2}(\frac{\partial}{\partial x} - i\frac{\partial}{\partial y})$ (13) holds iff $\Sigma\,(\frac{\partial f^\alpha}{\partial x})^2 = \Sigma\,(\frac{\partial f^\alpha}{\partial y})^2$ and $\Sigma\,\frac{\partial f^\alpha}{\partial x}\frac{\partial f^\alpha}{\partial y} = 0$. But these equations express precisely the condition that (x, y) be isothermal relative to the induced metric $f^*ds_E^2$.

Since $z = x + iy$ this is so iff f is conformal. Note that if we write $ds^2 = f^* ds_E^2 = h\, dz \cdot d\bar{z} = h(dx^2 + dy^2)$ as usual then $h = \Sigma\, (\frac{\partial f^\alpha}{\partial x})^2 = 2\, \Sigma\, |\eta^\alpha|^2 > 0$. Finally $\partial(\partial f^\alpha/\partial z)/\partial\bar{z} = \partial\eta^\alpha/\partial\bar{z} = 0$ since $\zeta^\alpha = \eta^\alpha dz$ are holomorphic. But (1) says that f is minimal iff $\Delta f^\alpha = 0$ iff $\partial^2 f^\alpha/\partial\bar{z}\partial z = 0$ for every α. This finishes the proof. □

Notice that if M is simply connected then the condition (14) of the above theorem automatically holds. Thus even in the absence of (14) we have a conformal minimal immersion defined on the universal cover of M, given (ζ^α) satisfying (12) and (13).

Let $\zeta = (\zeta^\alpha)$ be a collection of holomorphic 1–forms satisfying conditions (12) and (13) on a simply connected Riemann surface M. Define maps

$$f_\theta\colon M \to \mathbb{R}^n \ (0 \le \theta < \pi) \text{ by}$$

(16)
$$f_\theta\,(z) = 2\ \mathrm{Re} \int_{z_0}^{z} \zeta_\theta,$$

where $\zeta_\theta = (\zeta_\theta^\alpha) = (e^{i\theta}\zeta^\alpha)$. Then the collection (ζ_θ^α) satisfies (12) and (13) since $\Sigma\, |\eta_\theta^\alpha|^2 = \Sigma\, |\eta^\alpha|^2 > 0$ and $\Sigma\, (\eta_\theta^\alpha)^2 = e^{2i\theta}\, \Sigma\, (\eta^\alpha)^2 = 0$. Therefore f_θ are also conformal minimal immersions. From (10) we obtain

$$f_{\pi/2} = 2\ \mathrm{Re} \int i\zeta = 2\ \mathrm{Re} \int \tfrac{1}{2}\,(-{*}\, df + i\, df) \text{ and}$$

(17)
$$f_{\pi/2} = -\int *\, df$$

is called the *conjugate minmal surface to f.*

Notation. $f_{\pi/2} = \bar{f}$.

Given a conformal minimal immersion f: $M \to \mathbb{R}^n$, where we assume that M is simply connected (passing to the universal cover, if necessary), define a map

(18)
$$\Psi = \Psi_f = \frac{1}{\sqrt{2}}\,(\bar{f} + if)\colon M \to \mathbb{R}^n \oplus i\mathbb{R}^n = \mathbb{C}^n.$$

Observation. Ψ_f is holomorphic and $f^* ds^2_E = \Psi^* ds^2_{E\oplus E}$, where $ds^2_{E\oplus E}$ denotes the Euclidean metric on $\mathbb{C}^n$.

Ψ_f is called the associated holomorphic curve.

Let f: $M \to \mathbb{R}^n$ be a conformal (not necessarily minimal) immersion and also let $\zeta_f = (\zeta^\alpha)$ be defined as before. Locally $\zeta^\alpha = \eta^\alpha dz$, where $\eta^\alpha = \frac{\partial f^\alpha}{\partial z}$. (6) says that the map

(19) $$\Phi_f: M \to \mathbb{C}P^{n-1}, \quad z \mapsto [\eta^1(z), \cdots, \eta^n(z)]$$

is well–defined. In the above $[\eta^\alpha]$ denotes the line in $\mathbb{C}^n$ through 0 and (η^α). Φ_f is called the *Gauss map of f.*

Proposition. f is minimal if and only if Φ_f is holomorphic.

Proof. $\Delta f = 0$ iff $\partial(\partial f/\partial z)/\partial\bar{z} = 0$ iff $\partial\eta/\partial\bar{z} = 0$. □

More explicitly, the Gauss map of a conformal immersion f: $M \to \mathbb{R}^n$ is given by $\Phi_f(z) = [\frac{\partial f^\alpha}{\partial z}] = [\frac{1}{2}(\frac{\partial f^\alpha}{\partial x} - i\frac{\partial f^\alpha}{\partial y})] = [\frac{\partial f^\alpha}{\partial x} - i\frac{\partial f^\alpha}{\partial y}]$.

Recall that the hyperquadric in $\mathbb{C}P^{n-1}$, denoted by Q_{n-2}, is defined to be the algebraic variety given by

(20) $$(w^1)^2 + \cdots + (w^n)^2 = 0,$$

where (w^α) are the homogeneous coordinates on $\mathbb{C}P^{n-1}$.

(8) now says that

(21) $$\Phi_f(M) \subset Q_{n-2}.$$

(Remember that f is conformal iff $\Sigma\, (\eta^\alpha)^2 = 0$.)

The hyperquadric Q_{n-2} may be identifid with the Grassmann manifold of oriented two–planes in $\mathbb{R}^n$, denoted by $G_{n,2}$, as follows. Map $Q_{n-2} \to G_{n,2}$ by

(22) $$[w^1, \cdots, w^n] \mapsto [\mathrm{Re}(w^\alpha) \wedge \mathrm{Im}(w^\alpha)],$$

where $[\mathrm{Re}(w^\alpha) \wedge \mathrm{Im}(w^\alpha)]$ denotes the oriented two–plane in $\mathbb{R}^n$ spanned by the ordered pair of vectors $\{\mathrm{Re}(w^\alpha), \mathrm{Im}(w^\alpha)\}$.

$G_{n,2}$ can also be realized as the homogeneous space $SO(n)/SO(2)\times SO(2n-2)$: Take $A \in SO(n)$ and $[v_1 \wedge v_2] \in G_{n,2}$ $(v_i \in \mathbb{R}^n)$. Then $A\cdot[v_1 \wedge v_2]$ is simply $[Av_1 \wedge Av_2]$. This action is clearly transitive and the isotropy subgroup at $[E_1 \wedge E_2]$ ((E_α) = the Euclidean basis of $\mathbb{R}^n$) is $SO(2)\times SO(n-2)$. Write $A = (A_1, \cdots, A_n) \in SO(n)$. There is the projection map π: $SO(n) \to G_{n,2}$ given by $A \mapsto [A_1 \wedge A_2]$. (22) can be rewritten as

(23)
$$G_{n,2} \to Q_{n-2},\quad [A_1 \wedge A_2] \mapsto [A_1 + iA_2].$$

The Gauss map Φ_f: $M \to G_{n,2}$ then looks like

(24)
$$\Phi_f(z) = [(\frac{\partial f^\alpha}{\partial x}) \wedge (\frac{-\partial f^\alpha}{\partial y})] = [(\frac{\partial f^\alpha}{\partial y}) \wedge (\frac{\partial f^\alpha}{\partial x})].$$

Or, in terms of an adapted frame $(e_i;e_a)$ along f,

(25)
$$\Phi_f(z) = [e_2(z) \wedge e_1(z)].$$

Lemma. Let f: $M \to \mathbb{R}^n$ be a conformal immersion. Then f is minimal if and only if $\omega_1^a + i\omega_2^a$ are all of type (0,1) on M, where (ω_β^α) are written relative to any adapted frame along f. (See §2 (6), (11) for the definition.)

Proof. $\omega_1^a + i\omega_2^a = S_{1i}^a\varphi^i + iS_{2i}^a\varphi^i = S_{12}^a(\varphi^2 + i\varphi^1) + S_{11}^a\varphi^1 + iS_{22}^a\varphi^2$. Now $S_{12}^a(\varphi^2 + i\varphi^1) = iS_{12}^a\bar{\varphi}$ is a type (0,1) form. So $\omega_1^a + i\omega_2^a$ is of type (0,1) iff $S_{11}^a\varphi^1 + S_{22}^a i\varphi^2$ is of type (0,1) iff $S_{11}^a\varphi^1 + S_{22}^a i\varphi^2$ is a multiple of $\bar{\varphi} = \varphi^1 - i\varphi^2$ iff $S_{11}^a + S_{22}^a = 0$ iff f is minimal. □

Let $(\Omega_\beta^\alpha) = A^{-1}dA$ denote the Maurer–Cartan form of $SO(n)$. The invariant complex structure (coinciding with the inherited complex structure coming from $G_{n,2} = Q_{n-2} \subset \mathbb{C}P^{n-1}$) on $G_{n,2}$ is defined by letting the forms $(\Omega_1^a + i\Omega_2^a)$,

$a \geq 3$, pull back to type (1,0) forms on $G_{n,2}$. Now a local lifting of Ψ_f looks like

$$s\colon z \mapsto (e_2(z), e_1(z); e_a(z)), \tag{26}$$

where $(e_1, e_2 ; e_a)$ form an adapted frame along f. It follows that $s^*(\Omega_1^a + i\Omega_2^a) = \omega_2^a + i\omega_1^a$ and the preceding lemma says that f is minimal iff $s^*(\Omega_1^a + i\Omega_2^a)$ are of type (1,0) iff Φ_f is holomorphic.

For the rest of this section we work with the case n = 3.

Let φ be a meromorphic function on a Riemann surface M and μ be a not identically zero holomorphic 1–form on M. We further require that whenever φ has a pole of order m at $p \in M$ μ has a zero of order 2m at p. μ cannot have other zeros for if it did, say at p, then $\zeta^1(p) = \zeta^2(p) = \zeta^3(p) = 0$ and p would be a branch point. Put

$$\begin{aligned} \zeta^1 &= \tfrac{1}{2}(1 - \varphi^2)\mu, \\ \zeta^2 &= \tfrac{i}{2}(1 + \varphi^2)\mu, \\ \zeta^3 &= \varphi\mu. \end{aligned} \tag{27}$$

The ζ^α's have no common zeros, hence the condition (12) is met. The condition (13) is also easily verified. Therefore the holomorphic 1–forms $\zeta = (\zeta^\alpha)$ given by (27) define, at least on the universal cover of M, a conformal minimal immersion into $\mathbb{R}^3$.

On the other hand let $f\colon M \to \mathbb{R}^3$ be a given conformal minimal immersion. Put as before $\zeta^\alpha = \eta^\alpha dz$ $(\eta^\alpha = \frac{\partial f^\alpha}{\partial z})$, where on the right hand side we use a local holomorphic coordinate z. Assume that $\zeta^1 - i\zeta^2$ is not identically zero. (Note that if $\zeta^1 - i\zeta^2 \equiv 0$ then $\zeta^3 \equiv 0$. Thus f^3 is a constant map and f is a horizontal plane. This case is easily avoided by applying a rotation.) We put

$$\begin{aligned} \mu &= \zeta^1 - i\zeta^2, \\ \varphi &= \frac{\zeta^3}{\mu}. \end{aligned} \tag{28}$$

We thus obtain a meromorphic function φ and a holomorphic 1–form μ on M satisfying the prescription that at a pole of φ of order m μ has a zero of order 2m. To see this note that $(\eta^1 - i\eta^2)\cdot(\eta^1 + i\eta^2) = -(\eta^3)^2$, $\eta^1 + i\eta^2 = -(\frac{d\mu}{dz}\cdot\varphi^2)$, and $\eta^1 + i\eta^2$ is holomorphic. The pair $\{\mu, \varphi\}$ in (28) is called the *Weierstrass representative of f*.

The induced metric on M of a conformal minimal immersion f: $M \to \mathbb{R}^3$ is given by, in terms of the Weierstrass representative,

(29) $$ds^2 = (1 + |\varphi|^2)^2|\eta|^2\, dz\cdot d\bar{z},$$

where $\mu = \eta dz$. To see this just recall that $ds^2 = h\ dz\cdot d\bar{z}$ and $h = 2\,\Sigma\,|\eta^\alpha|^2$.

We leave the proof of the following proposition as an exercise.

Proposition. Let f: $M \to \mathbb{R}^3$ be a conformal minimal immersion and also let $\{\mu, \varphi\}$ be its Weierstrass representative. Then

(30) $$\varphi = \pi\circ\Phi_f^\perp,$$

where π: $S^2 \subset \mathbb{R}^3 \to \mathbb{C} \cup \{\infty\}$ is the stereographic projection and $\Phi_f^\perp$: $M \to S^2$ is the normal Gauss map, i.e., $\Phi_f^\perp(z)$ = the unit normal vector perpendicular to $f_*(T_z(M))$.

§4. Algebraic Gauss Maps

Digression. U(n) acts transitively on $\mathbb{C}P^{n-1}$ as follows: Take $v \in \mathbb{C}^n\backslash\{0\}$ and let $[v] \in \mathbb{C}P^{n-1}$ denote the complex line through 0 and v. Then for $A \in U(n)$

$$A\cdot[v] = [Av].$$

The isotropy subgroup at ${}^t[1, 0, \cdots, 0]$ is $U(1) \times U(n-1)$ and $\mathbb{C}P^{n-1}$ becomes a homogeneous space $U(n)/U(1)\times U(n-1)$. Consider the hyperquadric $Q_{n-2} \subset \mathbb{C}P^{n-1}$.

Identifying Q_{n-2} with $G_{n,2}$ we see that SO(n) acts transitively on Q_{n-2}. Let $[v_1 + iv_2] \in Q_{n-2}$, $v_i \in \mathbb{R}^n$. For $B \in SO(n)$

$$B \cdot [v_1 + iv_2] = [Bv_1 + iBv_2].$$

The isotropy subgroup at ${}^t[1, i, 0, \cdots, 0]$ is $SO(2) \times SO(n-2)$. Note that ${}^t[1, i, 0, \cdots, 0]$ corresponds to the oriented two-plane $[E_1 \wedge E_2]$ ((E_α) = the Euclidean basis in $\mathbb{R}^n$) in $G_{n,2}$.

$$\begin{array}{ccc} SO(n) & \xrightarrow{i_1} & U(n) \\ \pi_2 \downarrow & & \downarrow \pi_1 \\ Q_{n-2} & \xrightarrow[i_2]{} & \mathbb{C}P^{n-1} \end{array}$$

In the diagram above we let i_1, i_2 denote inclusions and π_1, π_2, projections, where

$$\pi_1(A) = A \cdot {}^t[1, 0, \cdots, 0] = [A_1],\ A = (A_\alpha) \in U(n),$$
$$\pi_2(B) = B \cdot {}^t[1, i, 0, \cdots, 0] = [B_1 + iB_2],\ B = (B_\alpha) \in SO(n).$$

Let (Θ^α_β), $1 \le \alpha,\beta \le n$, denote the Maurer–Cartan form of U(n) and also let (Ω^α_β) denote the Maurer–Cartan form of SO(n). So $i_1^*\Theta = \Omega$. The normalized Fubini–Study metric (so that the sectional curvature = 4) on $\mathbb{C}P^{n-1}$ is given by the pullback of the Adjoint invariant symmetric product

$$\sum_{\beta \ge 2} \Theta^1_\beta \bar{\Theta}^1_\beta.$$

A little computation shows that (see, for example, [J–R–Y] p.128) the restriction to Q_{n-2} of the above Fubini–Study metric is given by the pullback of the Adjoint invariant symmetric product

$$\frac{1}{2} \sum_{\substack{1 \le i \le 2 \\ 3 \le a \le n}} (\Omega^a_i)^2 = \frac{1}{2} \sum (\Omega^a_1 + i\Omega^a_2)(\Omega^a_1 - i\Omega^a_2)$$

on SO(n).

Notation. ds^2_{FS} will denote either the normalized Fubini–Study metric on $\mathbb{C}P^{n-1}$ or its restriction to Q_{n-2} depending upon the context.

Let $f\colon M \to \mathbb{R}^n$ be a conformal minimal immersion. There is the Gauss map $\Phi_f\colon M \to Q_{n-2}$. On Q_{n-2} we use the metric ds^2_{FS}. The induced Gaussian curvature K on M is nonpositive and we define the *total curvature of f* to be

$$\tau_f = \int_M K\, dA \le 0, \tag{1}$$

where dA is the area element of $(M, f^* ds^2_E)$.

Lemma. Let $A(\Phi_f)$ denote the area of Φ_f, i.e.,

$$A(\Phi_f) = \int_M \Phi_f^*\, dvol_{FS},$$

where $dvol_{FS}$ = the volume element of (Q_{n-2}, ds^2_{FS}). Then

$$-\tau_f = 2\, A(\Phi_f). \tag{2}$$

Proof.

$$\begin{array}{ccc} & & SO(n) \\ & s \nearrow & \downarrow \pi \\ U \subset M & \xrightarrow[\Phi_f]{} & Q_{n-2} \end{array}$$

Have

$$\Phi_f^*\, ds^2_{FS} = \tfrac{1}{2} s^* \,\Sigma\, (\Omega^a_1 + i\Omega^a_2)(\Omega^a_1 - i\Omega^a_2)$$
$$= \tfrac{1}{2} \Sigma\, (\omega^a_2 + i\omega^a_1)(\omega^a_2 - i\omega^a_1) \quad \text{by §3 (26).}$$

So

$$\Phi_f^*\, dvol_{FS} = \tfrac{i}{2}\tfrac{1}{2}\, \Sigma\, (\omega^a_2 + i\omega^a_1) \wedge (\omega^a_2 - i\omega^a_1)$$
$$= \tfrac{1}{2} \Sigma\, \omega^a_2 \wedge \omega^a_1 = \tfrac{-1}{2} \Sigma \det S^a\, \varphi^1 \wedge \varphi^2.$$

But §2 (20) says $K = \Sigma \det S^a$. So

$$A(\Phi_f) = \int_M \Phi_f^*\, dvol_{FS} = \tfrac{-1}{2} \int_M K\, \varphi^1 \wedge \varphi^2 = \tfrac{-1}{2}\, \tau_f,$$

since $\varphi^1 \wedge \varphi^2$ is the area form of $(M, f^* ds^2_E)$. □

Definition. Let $f\colon M \to \mathbb{R}^n$ be a conformal minimal immersion and $\Phi_f\colon M \to Q_{n-2}$, its Gauss map. We say that the Gauss map Φ_f is algebraic if

i) M is biholomorphic to a compact Riemann surface M_g punctured at a finite number of points $p_1, \cdots, p_d$;

ii) Φ_f extends to a holomorphic map $\Phi\colon M_g \to \mathbb{C}P^{n-1}$. (Note that $\Phi(M_g)$ is then, by Chow's theorem, an algebraic curve.)

Maintaining the preceding notation we suppose that Gauss map of f is algebraic. An *end* of the minimal surface f is, by definition, $f(\Delta_i)$, where Δ_i is a sufficiently small punctured disc (relative to the induced metric) in M centered at p_i, $1 \le i \le d$.

Remark. Given a conformal minimal immersion $f\colon M \to \mathbb{R}^n$ recall that the associated holomorphic curve $\Psi_f\colon M \to \mathbb{C}^n$ is defined to be $\Psi_f = \frac{1}{\sqrt{2}}(\bar{f} + i\tilde{f})$. Call Ψ_f *algebraic* if

i) M is biholomorphic to $M_g \setminus \{p_1, \cdots, p_d\}$;

ii) Ψ_f can be extended to a holomorphic map $\Psi\colon M_g \to \mathbb{C}^n \cup \mathbb{C}P^{n-1}_\infty = \mathbb{C}P^n$, i.e., Ψ_f can be realized as the affine part of Ψ with respect to the choice of a projective hyperplane at infinity. It would be interesting to compare the two notions of algebraic minimal surfaces.

If $f, g\colon M \to \mathbb{R}^n$ are *isometric* conformal minimal immersions then Ψ_f and Ψ_g are also isometric. The metric rigidity property of holomorphic curves in $\mathbb{C}^n$ then implies that Ψ_f and Ψ_g are congruent to each other, i.e., they differ by a translation followed by a unitary transformation of $\mathbb{C}^n$. Calabi [C3] found all noncongruent minimal surfaces isometric to a given holomorphic curve in $\mathbb{C}^n$.

Returning to the study of algebraic Gauss maps we have

Proposition. Let f: $M \to \mathbb{R}^n$ be a conformal minimal immersion whose Gauss map is algebraic. Then

$$\text{(3)} \qquad \frac{1}{\pi} A(\Phi_f) = d,$$

where d denotes the degree of the algebraic curve $\Phi(M_g) \subset \mathbb{C}P^{n-1}$.

Proof. This is a special case of the Wirtinger theorem ([G–H] p. 31). □

As a consequence $\Phi(M_g)$ intersects every hyperplane (not containing $\Phi(M_g)$) of $\mathbb{C}P^{n-1}$ exactly d times counting multiplicity. A significance of (3) is that

$$\text{(4)} \qquad -\tau_f = 2\pi d \text{ for some nonnegative integer d,}$$

assuming that Φ_f is algebraic.

Proposition. Let f: $M \to \mathbb{R}^3$ be a conformal minimal immersion whose Gauss map is algebraic. Then

$$\text{(5)} \qquad -\tau_f = 4\pi k$$

for some nonnegative integer k.

Proof. We first compute the area of (Q_1, ds^2_{FS}). Have $Q_1 = SO(3)/SO(2)$. $ds^2_{FS} = \psi\bar{\psi}$, $\psi = \frac{1}{\sqrt{2}} s^*(\Omega^3_1 + i\Omega^3_2)$, where s is a local section of $SO(3) \to Q_1$. Using the Maurer–Cartan structure equations of SO(3) and the relations

$$d\psi = -i(\text{the Levi–Civita connection form}) \wedge \psi,$$

$$i d(\text{the Levi–Civita connection form}) = \frac{K}{2} \psi \wedge \bar{\psi},$$

we compute that $K \equiv 2$. It follows that Q_1 is isometric to a two–sphere of radius $\frac{1}{\sqrt{2}}$ and the area of (Q_1, ds^2_{FS}) is $4\pi(\frac{1}{\sqrt{2}})^2 = 2\pi$. Now we have the extended Gauss map Ψ: $M_g \to Q_1 \subset \mathbb{C}P^2$ and $A(\Psi)$ (which equals $A(\Psi_f)$) is an integral multiple of the area of Q_1. (This is a special case of the equidistribution pro–

perty of holomorphic maps between compact Riemann surfaces.) □

Digression. Let M be a noncompact Riemann surface. A function $g \in C^2(M)$ is said to be subharmonic if $\Delta g \geq 0$, where Δ is the Laplace–Beltrami operator with respect to a metric in the conformal class of M. (The notions of harmonicity and subharmonicity are well–defined on a Riemann surface since the conditions $\Delta g = 0$ and $\Delta g \geq 0$ do not depend on the choice of a particular metric in the conformal class of M.) M is called *hyperbolic* if it carries a negative nonconstant subharmonic function. Otherwise M is called *parabolic.* A simply connected hyperbolic Riemann surface is biholomorphic to an open disc in $\mathbb{C}$ and a simply connected parabolic Riemann surface is biholomorphic to $\mathbb{C}$. Let $z_0 \in M$. A function g on M is called a *Green's function at* z_0 if

i) g is harmonic in $M \backslash \{z_0\}$;

ii) $g > 0$ in $M \backslash \{z_0\}$;

iii) if z is a holomorphic coordinate centered at z_0 then $g(z) + \log |z|$ is harmonic in a neighborhood of z_0. It is a fundamental result that M is hyperbolic if and only if there exists a Green's function at every point of M. (See [F–K] pp. 167–168 for a proof.)

Recall that a Riemannian manifold (N, ds_N^2) is said to be *complete* if it is a complete metric space (i.e., every Cauchy sequence is convergent). It is a well–known theorem that (N, ds_N^2) is complete iff every bounded subset of M is relatively compact iff every geodesic can be extended for arbitrary large values of the arclength parameter.

Our goal for the remainder of this section is to prove the Chern–Osserman theorem which states that a complete minimal surface has finite total curvature iff its Gauss map is algebraic. We establish this result with a sequence of lemmas. Our proof follows [L3] pp. 130–134 closely.

Lemma 1. Let D be a region in $\mathbb{C}$ (i.e., it is an open and connected subset) and also let $ds^2 = h\, dz \cdot d\bar{z}$ be a complete Riemannian metric on D. Suppose there exists a harmonic function λ on D such that

(6) $$\log h \leq \lambda.$$

Then $D = \mathbb{C}$ or $D = \mathbb{C} \setminus \{p\}$ for some $p \in \mathbb{C}$.

Proof. Put $\tilde{h} = e^{\lambda} > 0$, λ as in (6). So $\tilde{h} \geq h$. Then $\tilde{h}\, dz \cdot d\bar{z}$ also defines a complete metric in D since the length of a curve in $(D, \tilde{h}\, dz \cdot d\bar{z})$ is at least that of the same curve in $(D, h\, dz \cdot d\bar{z})$. Let $\tilde{D}$ denote the universal cover of D. The lift of λ to $\tilde{D}$ is the real part of a holomorphic function Λ. Consider

$$w(z) = \int_{z_0}^{z} e^{\Lambda(\xi)} d\xi,$$

a function on $\tilde{D}$. $|\frac{dw}{dz}| = |e^{\Lambda}| = e^{\lambda} = \tilde{h}$. Thus the map $w\colon \tilde{D} \to \mathbb{C}$ has an inverse defined in a maximum disc $\Delta(R) = \{w\colon |w| < R\} \subset \mathbb{C}$. The completeness of $\tilde{h}\, dz \cdot d\bar{z}$ implies that $R = \infty$, i.e., w is bijective and $\tilde{D} = \mathbb{C}$. Consider the holomorphic covering map $\pi\colon \tilde{D} \to D \subset \mathbb{C}$. Picard's theorem says that either $\pi(\tilde{D}) = \mathbb{C}$ or it misses a single point $p \in \mathbb{C}$. □

Lemma 2. Let D be an annular region given by

$$D = \{z \in \mathbb{C}\colon 0 < r < |z| < R \leq \infty\}$$

and also let $ds^2 = h\, dz \cdot d\bar{z}$ be a metric on D such that

(7) $\log h \leq \lambda$ for some harmonic function λ on D;

(8) each path $\{z(t)\colon 0 \leq t < 1\}$ in D with $\lim_{t \to 1} |z(t)| = R$ has infinite length.

Then $R = \infty$.

Proof. Suppose $R < \infty$. Then applying a biholomorphic map $z \mapsto cz$, $c \in \mathbb{R}$ we may assume that $r < \frac{1}{R} < 1 < R$. Put $D' = \{z: \frac{1}{R} < |z| < R\}$ and note that the metric $\tilde{h}\ dz \cdot d\bar{z}$ with $\tilde{h}(z) = h(z)h(\frac{1}{z})$ is complete in D'. Now $\log \tilde{h}(z) \leq \lambda(z) + \lambda(\frac{1}{z})$. So by Lemma 1 either $D = \mathbb{C}$ or $D = \mathbb{C}\backslash\{\text{a point}\}$ arriving at a contradiction. □

Lemma 3. Let $D \subset \mathbb{C}$ be a hyperbolic region and also let $ds^2 = h\ dz \cdot d\bar{z}$ be a metric in D whose Gaussian curvature satisfies

$$K \leq 0, \tag{9}$$

$$\int_M |K| < \infty. \tag{10}$$

Then there exists a harmonic function λ on D such that $\log h \leq \lambda$.

Proof. Using §1 (13), (14) we can rewrite (9), (10) as

$$(\partial/\partial x^2 + \partial/\partial y^2) \log h \geq 0, \tag{11}$$

$$\int_M (\partial/\partial x^2 + \partial/\partial y^2) \log h\ dx \wedge dy < \infty, \tag{12}$$

where $z = x + iy$ is the Euclidean coordinate on $D \subset \mathbb{C}$. Since D is hyperbolic at an arbitrary point $\xi \in D$ there is a Green's function $g_\xi > 0$ defined on $D\backslash\{\xi\}$ such that $G(z)$ is harmonic throughout D, where

$$G(z) = g_\xi(z) + \log |z - \xi|.$$

Set

$$F(z) = \frac{1}{2\pi} \int_M g_\xi(z)\ (\partial/\partial x^2 + \partial/\partial y^2) \log h\ dx \wedge dy.$$

$F(z)$ makes sense by virtue of (12) and moreover, $F(z) \geq 0$ by (11). Poisson's formula ([A] p. 166) gives

$$(\partial/\partial x^2 + \partial/\partial y^2)\ F = -(\partial/\partial x^2 + \partial/\partial y^2) \log h.$$

It follows that $F + \log h$ is harmonic in D. Now $F + \log h \geq \log h$ since $F \geq 0$. Take $\lambda = F + \log h$. □

Lemma 4. Let M be a complete Riemannian two–manifold such that

$$K \le 0, \tag{13}$$

$$\int_M |K| < \infty. \tag{14}$$

Then M is biholomorphic to a compact Riemann surface with finitely many points removed.

Proof. (14) implies that M is finitely connected ($\pi_1(M)$ is finitely generated) by a theorem of Huber [Hu]. This means we can find a relatively compact region $M_0 \subset M$ bounded by a finite number of regular Jordan curves $\Gamma_1, \cdots, \Gamma_d$ such that each component M_j of $M \backslash M_0$ can be biholomorphically mapped onto the annulus

$$D_j = \{z \in \mathbb{C}\colon 1 < |z| < r_j\},$$

where $\Gamma_j \leftrightarrow |z| = 1$. The region D_j is hyperbolic since $-\mathrm{Re}\,\{1 - \frac{1}{z}\} < 0$. The metric on D_j induced from M_j satisfies the conditions (9), (10). Lemma 2 then gives $r_j = \infty$. Put $\bar{D}_j = D_j \cup \{\infty\} \subset \mathbb{C} \cup \{\infty\} = \mathbb{C}P^1$. Then using maps $M_j \leftrightarrow D_j$ we can biholomorphically attach $\bar{D}_j$ to M and produce a compact Riemann surface $M_g \supset M$ with $M_g \backslash \{p_1, \cdots, p_d\} \cong M$. □

We can now prove

Theorem (Chern–Osserman [C–O1]). Let $f\colon M \to \mathbb{R}^n$ be a conformal minimal immersion. We further assume that M is complete with the induced metric. Then $-\tau_f < \infty$ if and only if the Gauss map Φ_f is algebraic.

Proof. Suppose Φ_f is algebraic. Then $-\tau_f = 2\pi d < \infty$ by (4). Conversely we suppose that $-\tau_f < \infty$. Then by Lemma 4 M is biholomorphic to $M_g \backslash \{p_1, \cdots, p_d\}$, where M_g is a compact Riemann surface of genus g. Let

$$\Delta_j = \{z \in \mathbb{C}\colon |z| < 1\}$$

be a local holomorphic coordinate system for M_g centered at p_j. In $\Delta_j\backslash\{0\}$ we have $\Phi_f(z) = [\frac{\partial f^1}{\partial z}, \cdots, \frac{\partial f^n}{\partial z}]$. It now suffices to show that the functions $\frac{\partial f^\alpha}{\partial z}$ have at most a pole at 0: If some of the functions $(\frac{\partial f^\alpha}{\partial z})$ had a pole at 0 then one simply replaces $(\frac{\partial f^\alpha}{\partial z})$ by $(z^k \frac{\partial f^\alpha}{\partial z})$, k = an integer bigger than the maximum order of the pole, thereby "removing" the pole. We see that with only poles to worry about $\Phi_f(z)$ extends to all of M_g. Assume that one of the $\frac{\partial f^\alpha}{\partial z}$'s had an essential singularity at $z = 0$. Then for almost all $v = (v^\alpha) \in \mathbb{C}^n$ the function $\Sigma\, v^\alpha \frac{\partial f^\alpha}{\partial z}$ would have an essential singularity at $z = 0$. Hence for almost all $v \in \mathbb{C}^n$ the function $\Sigma\, v^\alpha \frac{\partial f^\alpha}{\partial z}$ would take on the value 0 infinitely many times in every neighborhood of 0. Consider the hyperplane $H_v \cong \mathbb{C}P^{n-2}$ in $\mathbb{C}P^{n-1}$ given as the zero locus of the equation $\Sigma\, v^\alpha w^\alpha = 0$, where (w^α) are the homogeneous coordinates on $\mathbb{C}P^{n-1}$. Now $p \in H_v \cap \Phi_f(M)$ iff $\Sigma\, v^\alpha \frac{\partial f^\alpha}{\partial z}(p) = 0$. It follows that $\Phi_f(M)$ intersects a generic hyperplane in $\mathbb{C}P^{n-1}$ infinitely many times and (4) says that $-\tau_f = \infty$, contrary to our assumption. □

It follows at once that for a complete minimal surface in $\mathbb{R}^n$, without any assumption on the total curvature,

$$-\tau_f = 2\pi d, \text{ where } d \in \mathbb{Z}^+ \cup \{0\} \cup \{\infty\}. \tag{15}$$

For a complete minimal surface in $\mathbb{R}^3$

$$-\tau_f = 4\pi k, \text{ where } k \in \mathbb{Z}^+ \cup \{0\} \cup \{\infty\}. \tag{16}$$

Discussion. Let $f\colon M \to \mathbb{R}^n$ be a complete minimal surface with $-\tau_f < \infty$. Identify M with $M_g\backslash\{p_1, \cdots, p_d\}$ and let $\Delta_j = \{z \in \mathbb{C}\colon |z| < 1\}$ be a local coordinate system centered at p_j. In $\Delta_j\backslash\{0\}$ write $\Phi_f(z) = [\frac{\partial f^\alpha}{\partial z}]$. Since at 0 we have at worst a pole we can write, near 0,

$$\Sigma \left|\frac{\partial f^\alpha}{\partial z}\right|^2 = \frac{c}{|z|^{2m}} + \text{higher order terms},$$

for $c > 0$ and m, an integer. Now $ds^2 = 2\,\Sigma\,\left|\frac{\partial f^\alpha}{\partial z}\right|^2 dz\cdot d\bar{z}$ is complete in $\Delta_j \backslash \{0\}$ and this means a path approaching $p_j \leftrightarrow 0$ has an infinite length. It follows that we must have $m \geq 1$. In fact we must have

(17) $$m \geq 2.$$

If $m = 1$ then for suitable constants $0 \neq (c^\alpha) \in \mathbb{C}^n$ $\psi^\alpha = \frac{\partial f^\alpha}{\partial z} - \frac{c^\alpha}{z}$ would be holomorphic at zero. Thus

$$\mathrm{Re}\ c^\alpha \log z = \mathrm{Re}\int^z \left(\frac{\partial f^\alpha}{\partial z} - \psi^\alpha\right) dz = f^\alpha - \mathrm{Re}\int^z \psi^\alpha dz$$

would be a well–defined harmonic function at $z = 0$. So each c^α must be real. But $\Sigma\,(c^\alpha)^2 = 0$ and this would say $c^\alpha = 0$ for every α.

Digression. A canonical divisor on a compact Riemann surface M_g (g = the genus) is , by definition, the divisor of a meromorphic 1–form on M_g. Applying so called the Riemann–Hurwitz relation one can show that the degree of any canonical divisor is equal to $2g - 2$. (See [Y5] Chapter IV §4 for a proof.) In other words given a meromorphic 1–form on M_g the total number of zeros counted with multiplicity minus the total mumber of poles counted with multiplicity equals $2g - 2$.

Proposition. Let $f\colon M \to \mathbb{R}^n$ be a complete minimal surface with finite total curvature. Also let d denote the number of ends and g, the genus. Then

(18) $$\tau_f \leq 4\pi(1 - g - d).$$

Proof. Identify M with $M_g \backslash \{p_1, \cdots, p_d\}$ and note that each $\zeta^\alpha = \frac{\partial f^\alpha}{\partial z} dz$ gives a meromorphic 1–form on M_g. Put m_j = the maximum order of the poles of (ζ^α)

at p_j, $1 \le j \le d$. Picking suitable constants $(c^\alpha) \in \mathbb{C}^n$ the meromorphic 1–form

$$\eta = \Sigma\, c^\alpha \zeta^\alpha$$

has a pole of order exactly m_j at each p_j. Have $2g - 2 =$ (the number of zeros of η) – (the number of poles of η). So

$$\text{the number of zeros of } \eta = \Sigma\, m_j + 2g - 2 \ge 2g - 2 + 2d$$

by (17). But

$$-\tau_f = 2\pi\cdot(\text{the number of zeros of } \eta)$$

since

$$\text{the number of zeros of } \eta = |\Phi(M_g) \cap H_c|,$$

where $H_c \cong \mathbb{C}P^{n-2}$ is the hyperplane given by the equation $\Sigma\, c^\alpha w^\alpha = 0$ as in the proof of the Chern–Osserman theorem. □

Remark. Suppose f: $M \to \mathbb{R}^n$ is a complete minimal surface that is *embedded.* Then we must have equality in (18), i.e.,

(19) $$\tau_f = 4\pi\cdot(1 - g - d).$$

For a proof see [J–M].

§5. Examples

Example 1. The plane $\subset \mathbb{R}^3$ is a complete minimal embedded surface with total Gaussian curvature equal to 0. By (19) it has one end. Indeed $\mathbb{R}^2 \cong \mathbb{C}$ is biholomorphic to $\mathbb{C}P^1 = \mathbb{C} \cup \{\infty\}$ with one puncture.

Example 2. The Catenoid in $\mathbb{R}^3$ is given by the Weierstrass representative

$$\{\mu = \frac{1}{z^2}\, dz,\ \varphi(z) = z\}$$

on $M = \mathbb{C}\backslash\{0\}$, where z is the Euclidean coordinate. An explicit parametrization of the Catenoid is

$$(u, v) \mapsto (\sinh^{-1}u, (1 + u^2)^{1/2}\sin v, (1 + u^2)^{1/2}\cos v).$$

It is a surface of revolution obtained by revolving the Catenary $x^3 = \cosh x^1$ about the x^1–axis. From the above parametrization one computes the total curvature which is -4π. The surface is easily seen to be embedded with two ends. In fact Shoen [S] has shown that a complete minimal surface in $\mathbb{R}^3$ of finite total curvature of any genus with two ends must be the Catenoid.

Example 3. The Helicoid is given parametrically by

$$\mathbb{R}^2 \to \mathbb{R}^3; \ (u, v) \mapsto (u\cos v, u\sin v, v).$$

It is the conjugate surface to the Catenoid, hence locally isometric to the Catenoid. The Helicoid is a complete embedded ruled minimal surface of infinite total curvature.

Example 4. Take $M = \mathbb{C}$, $\mu = dz$, $\varphi(z) = z$. The resulting minimal surface in $\mathbb{R}^3$ is called Enneper's surface. For $z \in \mathbb{C} = M$, $(x^\alpha) \in \mathbb{R}^3$ it is given by

$$x^1 = \mathrm{Re}\,\{z - \tfrac{1}{3}z^3\},$$
$$x^2 = \mathrm{Re}\,\{i(z + \tfrac{1}{3}z^3)\},$$
$$x^3 = \mathrm{Re}\,\{z^2\}.$$

Enneper's surface is a complete minimal surface of total curvature -4π. It has genus 0 with one end. Note that this surface is not embedded.

Osserman [O2] proved that a complete minimal surface with $\tau_f = -4\pi$ in $\mathbb{R}^3$ is either the Catenoid (two ends) or Enneper's surface (one end). Chen [Che2] proved that a complete minimal surface with $\tau_f = -4\pi$ in any $\mathbb{R}^n$ is biholomorphic to $\mathbb{C}$ or $\mathbb{C}\backslash\{\text{a point}\}$, i.e., it has one or two ends of genus 0. It would be interesting to classify, up to congruence, complete minimal surfaces in $\mathbb{R}^n$ with total curvature -4π.

Example 5. Scherk's surface is given by the equation

$$\exp(x^3)\cos x^1 = \cos x^2$$

in the $x^1x^2x^3$-space. It is a complete embedded doubly periodic minimal surface of infinite total curvature.

Example 6. Let $\Lambda = \mathbb{Z} \otimes i\mathbb{Z}$ denote the integral lattice in $\mathbb{C}$. Requiring the projection π: $\mathbb{C} \to \mathbb{C}/\Lambda$ be holomorphic $M = \mathbb{C}/\Lambda$ becomes a Riemann surface, called a complex torus. $\wp(z)$ denote the Weierstrass function with respect to Λ, i.e.,

$$\wp(z) = \frac{1}{z^2} + \Sigma\left[\frac{1}{(z-w)^2} - \frac{1}{w^2}\right],$$

where the sum is taken over all $w \in \Lambda\backslash\{0\}$. $\wp(z)$, a meromorphic function on $\mathbb{C}$, is an elliptic function with periods in Λ. $\wp(z)$ has a double pole at each $w \in \Lambda$ with the principal part $\frac{1}{(z-w)^2}$ and is holomorphic elsewhere. $\wp(z)$ projects down to M and give a meromorphic function on M. We again use $\wp$ to denote this function. At the same time we confuse z with $\pi(z)$ and let the context dictate the proper meaning. Costa's surface [Co] is given by the Weierstrass representative

$$\{\mu = \wp(z)dz,\ \varphi(z) = 2(2\pi)^{1/2}\wp(\tfrac{1}{2})/\wp'(z)\}$$

on $M\backslash\{0, \frac{1}{2}, \frac{i}{2}\}$. The total curvature of Costa's surface is -12π. The surface is of genus one with three ends. Hoffman and Meeks [H–M2] showed that Costa's surface is actually embedded.

Remark. [H–M1] has shown, generalizing Example 6, the following: Let $g \geq 1$. Then there exists a complete embedded minimal surface f: $M \to \mathbb{R}^3$ of total curvature $-4\pi(g + 2)$, where $M = M_g\backslash\{\text{three points}\}$. It is in general an interesting problem to determine what restrictions one has on the number of ends of a complete minimal surface. For example, it is not known whether or not there exists a complete embedded minimal surface in $\mathbb{R}^3$ of genus 0 with finite total curvature other than the plane or the Catenoid.

Let F: $M \to \mathbb{C}^n$ be a holomorphic immersion from a Riemann surface M. The realification of F, denoted by $F_{\mathbb{R}}$, is defined to be the map

$$F_{\mathbb{R}} = \begin{bmatrix} \text{Re } F \\ \text{Im } F \end{bmatrix} : M \to \mathbb{R}^{2n},$$

i.e., $F_{\mathbb{R}}$ comes from F via the identification $\mathbb{R}^n \oplus i\mathbb{R}^n = \mathbb{C}^n$.

Lemma. $F_{\mathbb{R}}$ is a conformal minimal immersion.

Proof. F is an immersion So

(1) $$\text{the vector } \left(\frac{\partial F^{\alpha}}{\partial z}\right) \text{ is never zero.}$$

F holomorphic says

(2) $$\frac{\partial}{\partial x} \text{Re } F^{\alpha} = \frac{\partial}{\partial y} \text{Im } F^{\alpha}; \ \frac{\partial}{\partial x} \text{Im } F^{\alpha} = \frac{-\partial}{\partial y} \text{Re } F^{\alpha}.$$

(2) are the Cauchy–Riemann equations. Suppose $F_{\mathbb{R}}$ had a branch point, say at $z_0 \in M$. This is so iff

(3) $$\frac{\partial}{\partial z} \text{Re } F(z_0) = \frac{\partial}{\partial z} \text{Im } F(z_0) = 0.$$

But (3) together with (2) violates (1). This show that $F_{\mathbb{R}}$ is an immersion. Put $f^{\alpha} = \text{Re } F^{\alpha}$ and $f^{\alpha+n} = \text{Im } F^{\alpha}$. Then from (2) we obtain

$$\Sigma\left(\frac{\partial f^{\alpha}}{\partial z}\right)^2 + \left(\frac{\partial f^{\alpha+n}}{\partial z}\right)^2 = 0$$

which says that $F_{\mathbb{R}} = (f^{\alpha}, f^{\alpha+n})$ is conformal. Now $F_{\mathbb{R}}$ being an immersion is minimal iff

$$\partial^2 F_{\mathbb{R}}/\partial z \partial \bar{z} = 0 \text{ iff } (\partial^2/\partial x^2 + \partial^2/\partial y^2)F_{\mathbb{R}} = 0.$$

But this follows at once from (2). □

Example 7. Let M_g denote a compact Riemann surface of genus g. As a consequence of the Riemann–Roch theorem one can holomorphically and linearly fully embed M_g into $\mathbb{C}P^3$ with arbitrary large degree. That is to say, there exists a holomorphic embedding

$$\tilde{F}: M_g \to \mathbb{C}P^3$$

such that the image does not lie in a lower dimensional projective subspace and the number of intersections between $\tilde{F}(M_g)$ and a generic hyperplane in $\mathbb{C}P^3$ is equal to d. Moreover, we can choose d to be arbitrarily large. Take a hyper–plane at infinity $\mathbb{C}P^2_\infty$ so that $|\tilde{F}(M_g) \cap \mathbb{C}P^2_\infty| = d$. Taking the affine part of $\tilde{F}$ we obtain

$$F: M_g\backslash\{d \text{ points}\} \to \mathbb{C}^3.$$

This in turn gives a conformal minimal embedding $f = F_{\mathbb{R}}$ into $\mathbb{R}^6$.

§6. Minimal Immersions of Punctured Compact Riemann Surfaces

Let M_g denote a compact Riemann surface of genus g. Given a divisor D $\in \mathrm{Div}(M_g)$ one defines

$$L(D) = \{G \in \mathcal{M}^*(M_g): (G) + D \geq 0\} \cup \{0\},$$

where $\mathcal{M}^*(M_g)$ denotes the space of not identically zero meromorphic functions on M, and $(G) = (G)_0 - (G)_\infty$ denotes the divisor of G. L(D) is a $\mathbb{C}$–vector space and the Riemann–Roch theorem states that

$$\dim L(D) = \deg D - g + 1 + \dim (Z - D),$$

where Z is any canonical divisor. A detailed account of the Riemann–Roch theorem and related results can be found in [Y5] Chapter IV.

The main result of this section is

Theorem ([Y6]). Let F_1 be any nonconstant meromorphic function on a compact Riemann surface M_g of genus g. Then there exists another meromorphic function F_2 on M_g such that $\{dF_1, F_2\}$ is the Weierstrass pair defining a complete conformal minimal immersion of finite total curvature

$$(1) \qquad f: M \to \mathbb{R}^3,$$

where $M = M_g \setminus \{\mathrm{supp}(F_1)_\infty \cup \mathrm{supp}(F_2)_\infty\}$.

Proof. Let $F_1 \in \mathcal{M}^*(M_g)$ with

$$(F_1)_\infty = \Sigma\, b_i p_i,\ 1 \le i \le n,\ p_i \in M_g.$$

Also put $d = \Sigma\, b_i$. d is the degree of the polar divisor of F_1. dF_1, a meromorphic 1–form on M_g, has poles of order $b_i + 1$ at p_i and no other poles. Put

$$(dF_1)_0 = \Sigma\, a_j q_j,\ 1 \le j \le m,\ q_j \in M_g.$$

Have $2g - 2 = \deg\,(dF_1)_0 \quad \deg\,(dF_1)_\infty$ since $(dF_1) = (dF_1)_0 - (dF_1)_\infty$ is a canonical divisor. It follows that

$$\Sigma\, a_j = (2g - 2) + \Sigma\, b_i + n,\ 1 \le j \le m,\ 1 \le i \le n.$$

Define a divisor $D \in \mathrm{Div}(M_g)$ by

$$D = \Sigma\, a_j q_j - \Sigma\, c_i p_i,$$

where $1 \le j \le m$, $1 \le i \le n$ and $\Sigma\, c_i = 3g - 2 + d + n$. We also pick the c_i's such that $c_i \ge b_i + 1$ for every i. Note that $\deg D = -g$. The Riemann–Roch theorem gives

$$\dim L(-D) = \deg(-D) - g + 1 + \dim L((dF_1) + D) \ge 1.$$

Pick a nonconstant meromorphic function $G \in L(-D)$ (this is possible since $\dim L(-D) \ge 1$). Have

$$(G)_0 = \Sigma\, \tilde{a}_j q_j + \Sigma\, \tilde{a}_{m+k} q_{m+k},\quad 1 \le j \le m,\ 1 \le k \le l,$$

$$(G)_\infty = \Sigma\, \tilde{c}_i p_i,\ 1 \le i \le n.$$

Notice that we must have

$$\tilde{c}_i \le c_i;\ \ \tilde{a}_j \ge a_j;\ \ \Sigma\, \tilde{a}_j + \Sigma\, \tilde{a}_{m+k} = \Sigma\, \tilde{c}_i.$$

The last condition reflects the fact that (G) is a principal divisor and the first two conditions says that $G \in L(-D)$. Define a meromorphic function F_2 on M_g by

$$F_2 = \sum_{\alpha=1}^{\lambda} \frac{c_\alpha}{G^\alpha}\ \ ((c_\alpha) \ne 0),$$

where $\lambda = 2(n + m + l - 1) + 4g + 1$. The c_α's are some complex constants to be chosen suitably later. Since $\mathrm{supp}(F_2)_\infty = \mathrm{supp}(G)_0$ we obtain

$$\mathrm{supp}(F_2)_\infty = \{q_1, \cdots, q_{m+l}\}.$$

We consider the meromorphic 1–forms $F_2 dF_1$ and $F_2^2 dF_1$ on M_g. Observe that

$$\{q_{m+1}, \cdots, q_{m+k}\} \subset \mathrm{supp}(F_2 dF_1)_\infty \subset \{q_1, \cdots, q_{m+l}; p_1, \cdots, p_n\},$$

$$\{q_1, \cdots, q_{m+l}\} \subset \mathrm{supp}(F_2^2 dF_1)_\infty \subset \{q_1, \cdots, q_{m+l}; p_1, \cdots, p_n\}.$$

We *claim* that we can choose (c_α), not all zero, such that $F_2 dF_1$ and $F_2^2 dF_1$ have no residues and no periods on M_g. Put

$$R_{i\alpha} = \text{the residue of } \frac{dF_1}{G^\alpha} \text{ at } p_i,$$

$$R_{j\alpha} = \text{the residue of } \frac{dF_1}{G^\alpha} \text{ at } q_j,$$

$$R_{k\alpha} = \text{the residue of } \frac{dF_1}{G^\alpha} \text{ at } q_{m+k}.$$

So the residue of $F_2 dF_1$ at p_i is $\sum_\alpha c_\alpha R_{i\alpha}$, etc. Thus $F_2 dF_1$ on M_g has no residues if and only if

(A) $$\sum_\alpha c_\alpha R_{i\alpha} = 0;\ \ \sum_\alpha c_\alpha R_{j\alpha} = 0;\ \ \sum_\alpha c_\alpha R_{k\alpha} = 0.$$

Now the total residue of any meromorphic 1–form on a compact Riemann surface must vanish ([Y5] Chapter IV §4). Hence

$$\sum_{i,\alpha} c_\alpha R_{i\alpha} + \sum_{j,\alpha} c_\alpha R_{j\alpha} + \sum_{k,\alpha} c_\alpha R_{k\alpha} = 0.$$

It follows that (A) represents a homogeneous linear system in (c_α) containing at most $n+m+\ell-1$ independent equations. Let $(e_1, \cdots, e_{2g})$ be 1–cycles representing a (canonical) homology basis of M_g and put

$$P_{a\alpha} = \int_{e_a} \frac{dF_1}{G^\alpha}; \quad 1 \le a \le 2g, \ 1 \le \alpha \le \lambda.$$

$P_{a\alpha}$ is the e_a–period of $\frac{dF_1}{G^\alpha}$. So the e_a–period of the meromorphic 1–form $F_2 dF_1$ is $\sum_\alpha c_\alpha P_{a\alpha}$ and $F_2 dF_1$ has no periods on M_g if and only if

(B) $$\sum_\alpha c_\alpha P_{a\alpha} = 0.$$

This gives a homogeneous linear system in (c_α) containing 2g equations. We now consider the meromorphic 1–form $F_2^2 dF_1$. the residue at p_i of $F_2^2 dF_1$ is

$$R_i(c_\alpha) = R_{i2} c_1^2 + R_{i4} c_2^2 + \cdots + R_{i,2\lambda} c_\lambda^2 + 2R_{i3} c_1 c_2 + \cdots + 2R_{i,2\lambda} c_{\lambda-1} c_\lambda,$$

where $R_{i,2\lambda}$ denotes the residue at p_i of $\frac{dF_1}{G^{2\lambda}}$, etc. Thus $F_2^2 dF_1$ has no residues if and only if

(C) $$R_i(c_\alpha) = 0; \quad R_j(c_\alpha) = 0; \quad R_k(c_\alpha) = 0.$$

Again we can eliminate one of the equations from (C) using the fact that the total residue of $F_2^2 dF_1$ must vanish. Hence (C) represents a homogeneous quadratic system (R_i, R_j, R_k are all homogeneous polynomials in the c_α's of degree 2) containing $n+m+\ell-1$ equations. Requiring $F_2^2 dF_1$ to have no periods on M_g we obtain another homogeneous quadratic system (D) in (c_α) with 2g equations. The total number of equations in (A–D) is

$$2(n + m + \ell - 1) + 4g = \lambda - 1$$

and the *claim* follows. (Observe that in solving the system (A–D) we are intersecting a set of hyperplanes and homogeneous hyperquadrics in $\mathbb{C}^\lambda$.) Choose

(c_α) as in the above claim and put $M = M_g \backslash \Sigma$, where

$$\Sigma = \operatorname{supp}(F_1)_\infty \cup \operatorname{supp}(F_2)_\infty = \{p_1, \cdots, p_n; q_1, \cdots, q_{m+l}\}.$$

Then dF_1 is a holomorphic 1–form on M. Put

$$\text{(2)} \qquad \begin{aligned} \zeta^1 &= \tfrac{1}{2}(1 - F_2^2)dF_1, \\ \zeta^2 &= \tfrac{i}{2}(1 + F_2^2)dF_1, \\ \zeta^3 &= F_2 dF_1. \end{aligned}$$

The absence of periods and residues of $F_2 dF_1$, $F_2^2 dF_1$ on M_g guarantees that ζ^1, ζ^2, ζ^3 have no real periods on M. The Gauss map of f: $M \to \mathbb{R}^3$ arising from (ζ^α) extends holomorphically to all of M_g since the ζ^α's have at worst poles at the points of Σ. Let z denote a local holomorphic coordinate on M_g centered at p_i. dF_1 has a pole of order $b_i + 1$ at p_i. It follows that

$$h(z) = 2\,\Sigma\,|\eta^\alpha|^2 = \frac{c}{|z|^{2m}} + \text{higher order terms},$$

where $m \geq 2$ since $b_i \geq 1$. This shows that a path approaching p_i has infinite arclength. Similarly one checks that any other divergent path must have infinite arclength thus establishing that the induced metric on M is complete. □

An interesting question is to describe the space of non–congruent f's with $\mu = dF_1$, where F_1 is a given meromorphic function on M_g.

Observe that in the above proof

$$n \leq d; \quad m + l \leq 3g + d + n - 2.$$

Let # denote the total number of punctures of f, i.e., # is the cardinality of Σ. We then obtain

$$\text{(3)} \qquad \# = n + m + l \leq 3g + 3d - 2.$$

In fact we can say more. But first

Lemma. Let M_g be a compact Riemann surface of genus $g \geq 1$ and $p_1 \in M_g$, an arbitrary point. Then there exists a meromorphic function F whose polar divisor is dp_1, where $2 \leq d \leq g + 1$.

Proof. By Riemann–Roch

$$\dim L((g+1)p_1) = g + 1 + 1 - g + \dim L(Z - (g+1)p_1) \geq 2.$$

So there exists a nonconstant meromorphic function whose polar divisor is less than or equal to $(g+1)p_1$. Since the genus is ≥ 1 we cannot have a meromorphic function on M_g with exactly one simple pole for this would imply that $M_g \cong \mathbb{C}P^1$. The result follows. □

Of course a compact Riemann surface of genus zero is biholomorphically identified with $\mathbb{C}P^1$ and on $\mathbb{C}P^1$ there exists a meromorphic function whose polar divisor is dp for any $d \in Z^+$, $p \in \mathbb{C}P^1$.

Proposition. Let $p_1 \in M_g$, an arbitrary point. Then there exists a complete conformal minimal immersion of finite total curvature $f: M_g \backslash \Sigma \to \mathbb{R}^3$, where

i) $p_1 \in \Sigma$;

ii) Σ contains at most $3g + d$ points, where d is as in Lemma.

Proof. By assumption we have a meromorphic function F with a single pole at p_1 of order d. So $n = 1$. Also

$$m + l \leq 3g + d + n - 2 = 3g + d - 1.$$

Therefore $\# = n + m + l \leq 3g + d$. □

Corollary. Let M_g be as in the above. Then

i) it can be completely conformally and minimally immersed in $\mathbb{R}^3$ with at most $4g + 1$ punctures;

ii) if M_g is hyperelliptic then it can be completely conformally and minimally immersed in $\mathbb{R}^3$ with at most $3g + 4$ punctures.

Proof. First observe that $M_0 \cong \mathbb{C}P^1$ can be so immersed with one puncture as the xy–plane in $\mathbb{R}^3$. The rest of i) follows from the preceding proposition. Indeed let $p_1 \in M_g$ be any non–Weierstrass point. Then there exists a meromorphic function whose polar divisor is $(g+1)p_1$. To prove ii) just recall that on a hyperelliptic Riemann surface there exists a meromorphic function whose polar divisor has degree 2. □

The first part of Corollary was first proved in [G–K]. Also see [C–G].

§7. The Bernstein–Osserman Theorem

Let $F: U \subset \mathbb{R}^2 \to \mathbb{R}$ be a C^2 function and also let $S \subset \mathbb{R}^3$ be the graph of F, i.e., $S = \{(u, v, F(u,v)) \in \mathbb{R}^3: (u, v) \in U\}$.

Exercise. S is a (non–parametric) minimal surface if and only if

$$(1) \qquad (1 + |F_v|^2)F_{uu} - 2 <F_u, F_v> F_{uv} + (1 + |F_u|^2)F_{vv} = 0.$$

The classical Bernstein theorem states: *if* $F: \mathbb{R}^2 \to \mathbb{R}$ *is a function whose graph is a minimal surface then F is linear, that is, S is planar.*

Remark. The above theorem fails in the higher codimension case in that S does not have to be planar. For example, given any entire function $w = w(z): \mathbb{C} \to \mathbb{C}$, $S = \{(z, w(z): z \in \mathbb{C}\} \subset \mathbb{C}^2 = \mathbb{R}^4$ is a minimal surface.

We now give a proof of the Bernstein theorem following Chern [Cher3].

Proof of the Bernstein theorem. Suppose $S = \{(u, v, F(u,v)\} \in \mathbb{R}^3: (u, v) \in \mathbb{R}^2\}$ is a minimal surface. Given local isothermal coordinates (x, y) on S we have from §1

$$\Delta = \frac{-1}{h}\left(\frac{\partial^2}{\partial x^2} + \frac{\partial^2}{\partial y^2}\right), \quad K = \frac{1}{2}\, \Delta \log h,$$

where $ds^2 = h(dx^2 + dy^2)$ is the induced metric on $S \subset \mathbb{R}^3$, Δ is the Laplace–Beltrami operator of (S, ds^2), and K is the Gaussian curvature of (S, ds^2). Put $J = (1 + F_u^2 + F_v^2)^{1/2}$. Then a standard calculation gives

$$K = \Delta \log(\frac{J}{J+1}).$$

On S introduce a new metric $d\tilde{s}^2 = (\frac{J+1}{J})^2 ds^2$. $d\tilde{s}^2$ is conformally equivalent to ds^2 and its Gaussian curvature is identically zero. Since $1 \le (\frac{J+1}{J}) \le 2$ and ds^2 is complete we see that $d\tilde{s}^2$ is also complete. It follows that $(S, d\tilde{s}^2)$ is isometric to the uv–plane with the flat metric $du^2 + dv^2$. Since $K \le 0$ we obtain

$$-(\frac{\partial}{\partial u^2} + \frac{\partial}{\partial v^2}) \log(\frac{J}{J+1}) \le 0.$$

Note that the Laplace–Beltrami operator of (S, ds^2) is a multiple of $(\frac{\partial}{\partial u^2} + \frac{\partial}{\partial v^2})$ since ds^2 and $d\tilde{s}^2$ are multiples of each other. Now the above inequality says the function $\log(\frac{J}{J+1})$ is a subharmonic negative function on the uv–plane. The parabolicity of the uv–plane then implies that $\log(\frac{J}{J+1})$ must be a constant. Hence $K \equiv 0$ and S is planar. □

Theorem (Osserman [O5]). Let $f: M \to \mathbb{R}^3$ be a complete conformal minimal immersion of finite total curvature. Then if the Gauss map omits more than 3 points of $\mathbb{C}P^1$, f(M) is a plane.

Proof. Identify M with $M_g \backslash \{p_1, \cdots, p_d\}$, where M_g a compact Riemann surface of genus g. We have $\{\mu = dF, \varphi\}$, the Weierstrass representation pair of f. φ extends to M_g giving a holomorphic map

$$\hat{\varphi}: M_g \to \mathbb{C}P^1, \quad \hat{\varphi}|_M = \varphi.$$

Applying a rotation to f(M) if necessary we may (and do) assume:

i) $\mathrm{supp}(\hat{\varphi})_\infty \cap \{p_1, \cdots, p_d\} = \emptyset$;

ii) $(\hat{\varphi})_\infty$ consists only of simple poles.

Put

$$m = \deg \hat{\varphi}, \ B = \text{the total branching number of } \hat{\varphi}.$$

Applying the Riemann–Hurwitz formula to $\hat{\varphi}$ we obtain

(2) $$g = -m + 1 - B/2, \text{ or } B = 2(g + m - 1).$$

We now look at the differential dF and see how extends to all of M_g. dF has double zeros at the poles of φ and no other zeros. Near p_i, one of the punctures, we have

$$|\eta|^2(1 + |\varphi|^2)^2 = \frac{c}{|z|^{2m_i}} + \text{higher order terms}$$

with $2 \leq m_i \leq \infty$, where z is a local holomorphic coordinate centered at p_i. Thus $\mu = dF$ extends to a meromorphic 1–form $\hat{\mu}$ on M_g with a pole of order at each p_i (and no other poles). So

$$\text{supp}(\hat{\mu})_\infty = \{p_1, \cdots, p_d\}, \; \text{ord}_{p_i} = m_i \geq 2,$$

$$\text{supp}(\hat{\mu})_\infty \cap \text{supp}(\hat{\varphi})_\infty = \emptyset.$$

The degree of the divisor of μ, (μ), is $2g - 2$ since (μ) is a canonical divisor. Hence

(3) $$2g - 2 = 2m - \sum_{i=1}^{d} m_i.$$

Since $m_i \geq 2$ we must have

(4) $$g - 1 + d \leq m.$$

Suppose φ misses the points $q_1, \cdots, q_k$ of $\mathbb{C}P^1$. Then $\varphi^{-1}(\{q_1,\cdots,q_k\}) \subset \{p_1,\cdots,p_d\}$. Each q_i has m preimages counting multiplicity. So

(5) $$km \leq \sum_{i=1}^{d} (1 + n_i) = d + \Sigma\, n_i,$$

where $1 + n_i$ $(n_i \geq 0)$ is the multiplicity of $\hat{\varphi}$ at p_i. Now $\Sigma\, n_i$ is the sum of branching numbers at $\{p_1, \cdots, p_d\}$, hence it does not exceed the total branching number B. It follows that

(6) $$km \leq d + B = d + 2(g + m - 1).$$

Adding the inequalities in (4) and (6) and rewriting we obtain

(7) $$1 - g \leq (3 - k)m.$$

The inequality in (4) says $d - m \leq 1 - g$. So $d - m \leq (3 - k)m$, or

$$d \leq (4 - k)m. \tag{8}$$

But since M is not compact $d \geq 1$, hence $k < 4$. □

Observation. Let $f: M \to \mathbb{R}^3$ be a complete conformal minimal immersion of finite total curvature and suppose that the Gauss map of f omits three points. Then

$$\text{i) } g \geq 1; \quad \text{ii) } -\tau_f \geq 12\pi.$$

Proof. $k = 3$ and (7) becomes $1 - g \leq 0$. This proves i). Since $\varphi^{-1}(\{q_1,\cdots,q_k\}) \subset \{p_1,\cdots,p_d\}$ we get $k = 3 \leq d$ and (8) gives $d \leq m$. Thus $-\tau_f = 4\pi m \geq 4\pi d \geq 4\pi k = 12\pi$. □

Osserman [O3] proved that the Gauss map of a nonplanar complete conformal immersion $M \to \mathbb{R}^3$ (possibly with infinite total curvature) misses at most a set of logarithmic capacity zero. (A closed set Σ in $\mathbb{C}P^1$ is said to have zero logarithmic capacity if $\mathbb{C}P^1\backslash\Sigma$ is not hyperbolic. See [A–S] for details.) Xavier [X1, X2] in 1981 gave a strong improvement over Osserman's theorem showing that the Gauss map of a complete nonplanar minimal surface in $\mathbb{R}^3$ can not omit more than six points. Finally in 1988 Fujimoto [F3] proved that

Theorem (Fujimoto). Let $f: M \to \mathbb{R}^3$ be a complete conformal minimal immersion that is not planar. Then its Gauss map can not omit more than four points.

In the following we will show that there exists a complete minimal surface in $\mathbb{R}^3$ whose Gauss map misses k points, with any $k \leq 4$. Thus the bound "four" in Fujimoto's theorem is sharp.

Examples. i) the Catenoid has $k = 2$, $m = 1$.

ii) Enneper's surface has $k = 2$, $m = 1$.

iii) the Helicoid has $k = 2$, $m = \infty$.

iv) Put $M = \mathbb{C}\backslash\{p_1, \cdots, p_d\}$, $\varphi(z) = z$, $\mu = dz / \Pi(z - p_i) = \eta dz$, where z is the Euclidean coordinate on $\mathbb{C}$. $\{\mu, \varphi\}$ then defines a minimal surface f: $\tilde{M} \to \mathbb{R}^3$, where $\tilde{M}$ is either M or its universal cover. The Gauss map of f misses d+1 points $\{\infty, p_1, \cdots, p_d\}$. We claim that f is complete if $d \leq 3$. To see this take a path Γ approaching either ∞ or one of the p_i's and consider the integral

$$\int_\Gamma (1 + |z|^2)(1 / \Pi|z - p_i|)\ |dz|.$$

This integral represents the arclength of Γ and is infinite since either the integrand goes to ∞ or is aymptotically equal to $1/|z|$. This example is due to Voss [V].

v) Let $M = \mathbb{C}$, $\varphi(z) = z + \frac{1}{z}$, $\mu(z) = z^2dz$. Then the Gauss map of the resulting complete minimal surface is surjective.

The following question is open: does there exist a complete conformal minimal immersion $M \to \mathbb{R}^3$ with finite total curvature whose Gauss map misses three points?

Chern [Cher2] (also Osserman [O4]) gave a Bernstein type result for complete minimal surfaces in $\mathbb{R}^n$: *Let f: M $\to$ $\mathbb{R}^n$ be a complete nonplanar conformal minimal immersion. Then the subset of* $\mathbb{C}P^{n-1*}$ *(the space of hyperplanes in* $\mathbb{C}P^{n-1}$*) meeting with* $\Phi(M) \subset \mathbb{C}P^{n-1}$ *is dense.* See [Fu1, Fu2] for further results along these lines.

Quite recently, X. Mo and R. Osserman [M–O] established that the Gauss map of a nonplanar complete minimal surface in $\mathbb{R}^3$ of infinite total curvature takes on every value infinitely often, with the possible exeption of four points. The following question seems to be still open: Let $M \subset \mathbb{R}^3$ be a nonplanar complete minimal surface of infinite total curvature. Does the Gauss map of M take on every value of its image infinitely often?

Chapter II. Compact Minimal Surfaces in S^n

A minimal surface in S^n "is" a minimal cone (see §5 for a precise statement of this) in $\mathbb{R}^{n+1}$. Thus the study of minimal surfaces in S^n is in some sense more restricted in scope than that of minimal surfaces in Euclidean space.

Calabi [C2] (also Chern [Cher4]) gave a classification of minimal two–spheres in S^{2m}. We give an exposition of this in §2 using the method of moving frames. The key in Calabi's classification is the observation that minimal two–spheres in S^{2m} are determined by holomorphic data. Indeed given a conformal minimal immersion f: $S^2 = \mathbb{C}P^1 \to S^{2m}$ we can associate a holomorphic curve $\mathbb{C}P^1 \to \mathbb{C}P^{2m}$ called the directrix curve of f, and moreover, the minimal surface can be recovered from its directrix curve. We will see in later chapters that Calabi's work has far reaching consequences in the study of minimal surfaces in more general spaces.

Recently Bryant [Br1] proved that any compact Riemann surface can be conformally and minimally immersed in S^4. His idea was to use the Penrose fibration $\mathbb{C}P^3 \to S^4$ and reduce the immersion problem to a problem in Algebraic Geometry. We give a detailed account of his theorem in §§3–4.

Lawson [L1] gave examples of compact minimal surfaces in S^3 of every genus. As a consequence an arbitrary compact Riemann surface can be minimally, not necessarily conformally, immersed in S^3. It would be very interesting to decide whether or not every compact Riemann surface can be conformally and minimally immersed into S^3.

§1. Moving Frames

Let $N^n = G/H$ be a homogeneous space of dimension n. G is a Lie group acting transitively on the smooth manifold N and H is a closed subgroup. H is the isotropy subgroup at a point $0 \in N$, i.e., $H = G_0 = \{g \in G: g\cdot 0 = 0\}$. Points of N may be thought of as left cosets in G/H.

$$\pi: G \to N, \quad g \mapsto gH = g\cdot 0.$$

Suppose we are given an immersion $f: M \to N$ from an m–dimensional manifold M.

Definition. A local section s of the pullback bundle $f^{-1}G \to M$ is called a moving frame along f. That is to say, s is a smooth map $U \subset M \to G$ such that $\pi \circ s = f$.

Let $\mathfrak{g}$ (respectively, $\mathfrak{h}$) denote the Lie algebra of G (respectively, H). Fix a subspace $\mathfrak{n}$ ($\mathfrak{n}$ is not a subalgebra in general) of $\mathfrak{g}$ complementary to $\mathfrak{h}$;

(1) $$\mathfrak{g} = \mathfrak{h} \oplus \mathfrak{n}.$$

There is the linear frame bundle $L(N) \to N$. For $p \in N$ the fibre $\pi^{-1}(p)$ consists of all frames of T_pN. L(N) is a $GL(n;\mathbb{R})$–principal bundle over N:

$$\text{for a frame } u = (u_i) \text{ and } A = (a^i_j) \in GL(n;\mathbb{R}),\ u\cdot A = (u_j a^j_i).$$

Choose a basis $E = \{E_1, \cdots, E_n\}$ of $\mathfrak{n}$ and consider the map

(2) $$i: G \to L(N), \quad g \mapsto g_{*0}(E_1, \cdots, E_n).$$

Via $\pi_{*e}: \mathfrak{g} \to T_0N$ we identify $\mathfrak{n}$ with T_0N.

(3) $$i|_H: H \to GL(\mathfrak{n})$$

is called the linear isotropy representation. Using the reference basis (E_i) of $\mathfrak{n}$ we have

$$i|_H: H \to GL(n;\mathbb{R}),$$

and

$$i(gh) = i(g)\cdot i|_H(h).$$

Examples. i) Let $G = E(n) = O(n;\mathbb{R})\cdot\mathbb{R}^n$ (semi–direct product), $H = O(n;\mathbb{R})$. More explicitly, $E(n)$, called the Euclidean group of rigid motions in $\mathbb{R}^n$, consists of all pairs (h, x), where h is an orthogonal transformation and x is a translation in $\mathbb{R}^n$. Have

$$(h_1, x_1)\cdot(h_2, x_2) = (h_1h_2, x_1 + h_1(x_2)),$$
$$(h, x)^{-1} = (h^{-1}, -h^{-1}(x)).$$

G/H is just $\mathbb{R}^n$. $H = O(n)$ is the isotropy subgroup at $0 = {}^t(0, \cdots, 0) \in \mathbb{R}^n$. Put

$$\mathfrak{n} = \{(0, y): y \in \mathbb{R}^n\}.$$

Have the vector space direct sum

$$\mathfrak{g} = \mathfrak{o}(n) \oplus \mathfrak{n}.$$

Put $E = (\epsilon_1, \cdots, \epsilon_n)$, the canonical basis of $\mathbb{R}^n$. We then have

$$i\colon E(n) \to L(\mathbb{R}^n),$$
$$(h, x) \mapsto (h, x)_{*0}(\epsilon_1, \cdots, \epsilon_n),$$

and $(h, x)_{*0}(\epsilon_i)$ is just the orthonormal frame at $x \in \mathbb{R}^n$ given by the column vectors of the matrix h. We see that i gives the identification

$$E(n) \leftrightarrow O(\mathbb{R}^n) \subset L(\mathbb{R}^n),$$

where $O(\mathbb{R}^n) \to \mathbb{R}^n$ is the bundle of orthonormal frames.

ii) Let $G = SO(n+1;\mathbb{R})$ and also let $H = SO(n;\mathbb{R})$, where $H \subset G$ via $h \mapsto \binom{1,0}{0,h}$. $G/H = S^n$, the standard n–sphere in $\mathbb{R}^{n+1}$. G acts on S^n by

$$g\colon v = {}^t(v^0, v^1, \cdots, v^n) \in S^n \mapsto g\cdot v,$$

and $SO(n) = G_0$, where $0 = {}^t(1, 0, \cdots, 0)$. For a complementary subspace to

$\mathfrak{g}_0 = \mathfrak{o}(n)$ we take

$$\mathfrak{n} = \left\{ \begin{bmatrix} 0, & {}^t y \\ y, & 0_n \end{bmatrix} : y \in \mathbb{R}^n,\ 0_n \text{ is the } n\times n \text{ zero matrix} \right\}.$$

We identify $\mathfrak{n}$ with $\mathbb{R}^n$ via $\begin{bmatrix} 0, & {}^t y \\ y, & 0_n \end{bmatrix} \mapsto y$ and with this identification in mind we take for a basis of $\mathfrak{n}$, $E = \{\epsilon_1, \cdots, \epsilon_n\}$, where $\epsilon_1 = {}^t(1, 0, \cdots, 0)$, etc. We have

$$i\colon SO(n+1) \to L(S^n),$$

$$g \mapsto g_{*0}(\epsilon_1, \cdots, \epsilon_n).$$

Now $g(0) = g\cdot 0$ is the first column vector of g.

Notation. $g = (g_0, g_1, \cdots, g_n) \in SO(n+1)$, $h = (0, h_i) = (h_i) \in SO(n)$.

For $h \in SO(n)$,

$$h\cdot(\epsilon_1, \cdots, \epsilon_n) = (h_1\epsilon_1, \cdots, h_n\epsilon_n)$$

and we see that $i(SO(n+1))$ is identified with the bundle of oriented orthonormal frames over S^n.

$$\begin{array}{c} SO(n+1) \cong SO(S^n) \\ \downarrow \\ S^n \end{array}$$

Coming back to the general discussion we let $f\colon M \to N = G/H$ be a smooth immersion and also let $e\colon U \subset M \to G$ be a moving frame along f. The Maurer–Cartan form of G, denoted by Ω, is a $\mathfrak{g}$-valued left-invariant 1-form on G defined by

(4) $$\Omega_g(X) = L_{g^{-1}*}X, \quad X \in T_gG.$$

Ω decomposes into

$$\Omega = \Omega_{\mathfrak{h}} \oplus \Omega_{\mathfrak{n}}, \quad \Omega_{\mathfrak{n}} = \Omega^\alpha \otimes E_\alpha,$$

where (E_α) form a basis of $\mathfrak{n}$. Given a coframe $\varphi = (\varphi^1, \cdots, \varphi^m)$ on $U \subset M$ we can write

$$e^*\Omega^\alpha = X_i^\alpha\, \varphi^i, \tag{5}$$

for some smooth functions (X_i^α): $U \to \mathbb{R}^{n\times m}$. The idea is to choose e so as to "maximally simplify" (X_i^α) and extract information about f by differentiating both sides of the equations in (5) using the Maurer–Cartan structure equations of the group G. [C] is the standard reference on the theory of moving frames. For a more modern treatment the reader may consult [Gr] or [J]. The case of compact G is discussed in detail in [Y3].

For the rest of this section we concentrate on the case $G = SO(n+1)$ and $H = SO(n)$. A local orthonormal frame in $S^n = G/H$ is a smooth map

$$s\colon U \subset S^n \to SO(n+1),$$

$$s(x) = (s_0(x), \cdots, s_n(x)),\ s_0(x) = x.$$

Recall from Example ii) the decomposition $\Omega = \Omega_{\mathfrak{h}} \oplus \Omega_{\mathfrak{n}}$ and

$$\Omega_{\mathfrak{n}} = \Omega_0^1 \otimes E_1 + \cdots + \Omega_0^n \otimes E_n.$$

There are the Maurer–Cartan structure equations of G given by

$$d\Omega_\beta^\alpha = -\Omega_\gamma^\alpha \wedge \Omega_\beta^\gamma, \tag{6}$$

where $\Omega = (\Omega_\beta^\alpha)$, $0 \le \alpha,\beta \le n$, is the $\mathfrak{o}(n+1)$–valued Maurer–Cartan form of G.

Remark. Let $X = (X_\beta^\alpha)$: $GL(n+1;\mathbb{R}) \to \mathbb{R}^{(n+1)^2}$ be the matrix coordinates. Writing $dX = (dX_\beta^\alpha)$ we get $\Omega = X^{-1}dX$ with the usual tangent space identification. The exterior differentiation of both sides of the equation $dX = X\Omega$ leads to $d\Omega = -\Omega \wedge \Omega$ which, written out in components, gives the Maurer–Cartan structure equations of $GL(n+1;\mathbb{R})$. For a closed subgroup $G < GL(n+1;\mathbb{R})$, say $G = SO(n+1)$, Ω and the structure equations are obtained simply by restrictions.

Notation. $s^*\Omega^\alpha_\beta = \omega^\alpha_\beta$.

Lemma. The symmetric product $(\Omega^1_0)^2 + \cdots + (\Omega^n_0)^2$ is Ad(H)–invariant.

Proof. This follows from the formula

$$\mathrm{Ad}(h)\cdot X = hXh^{-1}, \quad h \in H,\ X \in \mathfrak{g}. \quad \square$$

Proposition. $ds^2 = (\omega^1_0)^2 + \cdots + (\omega^n_0)^2$ is the standard metric on the unit sphere $S^n \subset \mathbb{R}^{n+1}$.

Proof. The identification $\mathfrak{n} = T_0S^n$ induces a bijective correspondence between the set of Ad(H)–invariant inner products in $\mathfrak{n}$ and that of invariant metrics on $S^n = G/H$. Any two invariant metrics on a Riemannian symmetric space are constant multiples of each other. By Lemma ds^2 is an invariant metric on S^n. It follows that $ds^2 = c\cdot$(the standard metric). Will see shortly that the sectional curvature of (S^n, ds^2) is $\equiv 1$, hence $c = 1$. $\square$

Notation. $0 \leq \alpha,\beta \leq n$, $1 \leq a,b,c \leq n$, $\omega^a_0 = \omega^a$.

Using (6) we compute that

$$d\omega^a = d\, s^*\Omega^a_0 = s^*\, d\Omega^a_0 = -\omega^a_\alpha \wedge \omega^\alpha_0.$$

Now $\omega^0_0 = 0$ since Ω is skew–symmetric (i.e., $\mathfrak{o}(n+1)$–valued). Thus we obtain

$$d\omega^a = -\omega^a_b \wedge \omega^b. \tag{7}$$

That is to say, $(s^*\Omega^a_b)$ are the Levi–Civita connection forms of (S^n, ds^2) with respect to the orthonormal frame s. Now

$$d\omega^a_b = -\omega^a_\alpha \wedge \omega^\alpha_b = -\omega^a \wedge \omega_b - \omega^a_c \wedge \omega^c_b.$$

So,

$$d\omega^a_b = -\omega^a_c \wedge \omega^c_b + \omega^a \wedge \omega^b. \tag{8}$$

The curvature forms (θ^a_b) are, by definition,

$$\theta^a_b \equiv d\omega^a_b + \omega^a_c \wedge \omega^c_b.$$

From (8) we see that

(9) $$\theta^a_b = 1\cdot\omega^a \wedge \omega^b$$

showing that (S^n, ds^2) is a space of constant sectional curvature 1.

§2. Minimal Two–Spheres in S^n

Let M be a Riemann surface and consider a conformal immersion

(1) $$f\colon M \to S^n.$$

Notation. ds^2_E = the standard metric on S^n, $ds^2 = f^* ds^2_E$.

Let Δ denote the Laplace–Beltrami operator of (M, ds^2).

Exercise. Let f be as in the above. Then

f is minimal if and only if $\Delta f + 2f = 0$.

Given f as in (1) take a moving frame along f,

(2) $$e = (e_0, \cdots, e_n)\colon U \subset M \to SO(n+1),\quad e_0 = f.$$

Definition. A Darboux frame along f is a moving frame e such that e_a is perpendicular to f (in $\mathbb{R}^{n+1}$) for $a \geq 3$, i.e., the tangent plane at f(x) to f is spanned by $e_1(x)$ and $e_2(x)$ for every $x \in U \subset M$.

Now $f^{-1}G = f^{-1}SO(S^n) \to M$ is a SO(n)–principal bundle over M and we see easily that there always exists a Darboux frame along f. For a Darboux frame e

(3) $$e^*\Omega^a_0 = \omega^a = 0 \quad \text{for } a \geq 3.$$

Also $ds^2 = e^*((\Omega^1_0)^2 + \cdots + (\Omega^n_0)^2) = (\omega^1)^2 + (\omega^2)^2$ and (ω^1, ω^2) form a (local) orthonormal coframe on M.

Index Convention. $0 \leq \alpha,\beta,\gamma \leq n$, $1 \leq i,j,k \leq 2$, $3 \leq a,b,c \leq n$.

Exterior differentiating both sides of the equations in (3) gives

$$0 = d\omega^a = -\omega_1^a \wedge \omega^1 - \omega_2^a \wedge \omega^2.$$

By Cartan's Lemma there exixt local functions $h_{ij}^a = h_{ji}^a$ such that

(4) $$\omega_i^a = h_{ij}^a \omega^j.$$

Hopf [Hop1] defines

(5) $$S^a = \frac{-1}{2}(h_{11}^a - h_{22}^a) + ih_{12}^a, \quad S = (S^a): U \subset M \to \mathbb{C}^{n-2}.$$

The second fundamental forms written relative to e are

(6) $$II^a = h_{ij}^a \omega^i \cdot \omega^j \text{ (symmetric product).}$$

Assume now that f is *minimal*, where we take the vanishing condition

(7) $$\text{trace } h^a = 0, \quad a \geq 3$$

to be the definition of minimality. (Note that trace h^a is independent of the choice e since the trace is preserved under the orthogonal change of basis.) As a consequence of our assumption we can write

$$S^a = -h_{11}^a + ih_{12}^a, \text{ or}$$

(8) $$S^a \varphi = -\omega_1^a + i\omega_2^a, \text{ where } \varphi = \omega^1 + i\omega^2.$$

Review. Consider a conformal immersion

$$f: M \to (N, ds_N^2),$$

where N^n is a Riemannian manifold. Pick a (local) orthonormal frame e in N and let $\omega = (\omega^1, \cdots, \omega^n)$ be the dual coframe. So, $ds_N^2 = (\omega^1)^2 + \cdots + (\omega^n)^2$. We say that e is a Darboux frame along f if in a sufficiently small neighborhood

$$f^*\omega^a = 0, \text{ or } e_a \perp f \quad \text{for } a \geq 3.$$

(By a Darboux frame along f one also means a map e: $U \subset M \to f^{-1}O(N)$ ($O(N) \to N$, the bundle of orthonormal frames over N) such that $e_a \perp f$. This confusion is in practice proven to be harmless since given such an e one can always extend it to a Darboux frame defined in a neighborhood in N and the computation taking place in M does not depend on the extension.) Given a Darboux frame we have

$$0 = d(f^*\omega^a) = -f^*\omega^a_i \wedge f^*\omega^i, \ a \geq 3, \ i = 1, 2,$$

where (ω^A_B), $1 \leq A,B \leq n$, are the Levi–Civita connection forms of (N, ds^2_N) relative to e. (Also $f^*\omega^1 + if^*\omega^2$ is of type (1,0) or of type (0,1) depending upon the orientation of e_1, e_2.) It follows that

$$f^*\omega^a_i = h^a_{ij} f^*\omega^j, \quad h^a_{ij} = h^a_{ji}, \text{ local functions on M.}$$

The second fundamental forms of f relative to e are

$$II^a = h^a_{ij} f^*\omega^i \cdot f^*\omega^j.$$

f is said to be minimal if trace h^a vanishes for every $a \geq 3$.

Coming back to our main discussion we have

Lemma A. $\Lambda = {}^tSS\varphi^4$ is a globally defined symmetric type (4,0) form on M.

Proof. Let $\tilde{e}$: $\tilde{U} \subset M \to SO(n+1)$ be another Darboux frame along f. Then $\tilde{e}$ and e are related by

$$\tilde{e} = e \cdot h, \tag{9}$$

where $h: U \cap \tilde{U} \to SO(2) \times SO(n-1)$. Write $h = \left(\begin{pmatrix} \cos t, & -\sin t \\ \sin t, & \cos t \end{pmatrix}, A\right)$. Define tilded quantities $\tilde{\omega}$ and $\tilde{S}$ using $\tilde{e}$. We then obtain

$$\tilde{\omega} = e^{it}\omega, \quad \tilde{S} = e^{2it}A^{-1}S \tag{10}$$

using the transformation rules

$$Ad(h)X = hXh^{-1}, \ X \in \mathfrak{n} \ (\mathfrak{n} \text{ is } Ad(H)\text{–invariant}),$$

$$\tilde{e}^*\Omega_{\mathfrak{n}} = Ad(h^{-1}) \, e^*\Omega_{\mathfrak{n}}.$$

It follows that $\tilde{\Lambda} = \Lambda$ and Λ is well–defined on M. The rest is easy. □

Lemma B. Λ is holomorphic.

Proof. Have $\varphi = \eta dz$, where z is a local holomorphic coordinate in M. Then $\Lambda = {}^tSS\eta^4(dz)^4$ and we will show that ${}^tSS\eta^4$ is holomorphic. This is so iff $\partial({}^tSS\eta^4)/\partial\bar{z} = 0$ iff $d({}^tSS\eta^4) \equiv 0$ (mod dz or mod φ). We have

(*) $$d\omega = i\omega_2^1 \wedge \varphi, \text{ hence}$$

(**) $$d\eta \equiv i\eta\omega_2^1 \text{ (mod dz)}.$$

Exterior differentiate both sides of the equations in (8) and use (*) to obtain

(†) $$(dS^a + 2i\, S^a\omega_2^1 + S^b\omega_b^a) \wedge \varphi = 0.$$

Thus we have

(***) $$d({}^tSS) \equiv -2i{}^tSS\omega_2^1 \text{ (mod dz)}.$$

The result follows from (**) and (***). □

We now *assume* that $M = \mathbb{C}P^1 \cong S^2$ and proceed to give a complete description of conformal minimal immersions $f: \mathbb{C}P^1 = S^2 \to S^n$. By the Riemann–Roch theorem we then must have

(11) $$\Lambda \equiv 0.$$

Condition (11) means that S is isotropic, i.e., ${}^tSS = 0$. In terms of its real and imaginary parts this is equivalent to saying

(12) $$\langle \mathrm{Re}\, S, \mathrm{Im}\, S\rangle = 0, \quad |\mathrm{Re}\, S| = |\mathrm{Im}\, S|,$$

where $\langle\ ,\ \rangle$ and $|\ ,\ |$ denote the standard inner product and norm in $\mathbb{R}^{n-2}$.

Define a function $\tau^2 (\tau \geq 0)$: $M \to \mathbb{R}$ by

(13) $$\tau^2 = |S|^2 = {}^t\bar{S}S.$$

The verification of the following is routine and left to the reader as an exercise.

Observation. i) Let K denote the Gaussian curvature of $(M = \mathbb{C}P^1, ds^2)$. Then

$$K = 1 - 2\tau^2.$$

In particular, τ^2 is a smooth function on M. τ is continuous everywhere and smooth away from the zeros.

ii) If $\tau \equiv 0$ then f(M) is totally geodesic.

iii) For f: $M = \mathbb{C}P^1 \to S^3$ we must have $\tau \equiv 0$, hence by ii) f(M) is an equator.

Hereafter we exclude the case $\tau \equiv 0$ from our discussion.

Definition. Let U be a domain in M. A smooth function h: $U \to \mathbb{C}$ is said to be of *analytic type* if for each $x \in U$, if z is a local holomorphic coordinate centered at x, then

$$h = z^b \hat{h},$$

where $b \in \mathbb{Z}^+$, $\hat{h}$ is a smooth function with $\hat{h}(0) \neq 0$. It is known [Cher4] that the functions of analytic type are exactly solutions of exterior equation

$$dh \equiv h\psi \ (\mathrm{mod}\ dz), \tag{14}$$

where ψ is a $\mathbb{C}$–valued 1–form on U. So if h is of analytic type on U then h is either identically zero or its zeros are isolated and of finite order (the integer b in the above definition is the order at x).

Lemma C. τ is of analytic type.

Proof. From (†) in the proof of Lemma B we get

$$dS^a \equiv 2i\ S^a \omega_1^2 - S^b \omega_b^a \ (\mathrm{mod}\ dz).$$

Or using the matrix notation

$$\partial S/\partial \bar{z} = XS \quad \text{for some local functions } X = (X_b^a).$$

(14) does the rest. □

Observation. $\tau(p) = 0$ if and only if $p \in M$ is an umbilic point.

Notation. Z = the zero set of τ.

Using the transformation rules given in (10) we see that in a neighborhood of any point $x \in M \backslash Z$ we can choose a Darboux frame e along f such that relative to e

(15) $$S = \tau^t(1, i, 0, \cdots, 0) \colon U \to \mathbb{C}^{n-2}.$$

We now look near a point $x \in Z = \text{Zero}(\tau)$. Since τ is of analytic type there exists a positive integer b with

$$S = z^b \hat{S},$$

where z is a local holomorphic coordinate centered at x and $\hat{S}(0) \neq 0$. It follows that near $x \in Z$ we can have relative to a suitable Darboux frame e

(16) $$S = z^b \hat{\tau}^t(1, i, 0, \cdots, 0) \colon U \to \mathbb{C}^{n-2},$$

where $\hat{\tau} = |\hat{S}|$. Summarizing what we have so far

Theorem 1. Let $f \colon M = \mathbb{C}P^1 \to S^n$ be a conformal minimal immersion that is not totally geodesic. Then there exists a Darboux frame along f with respect to which we can write near any point of $\mathbb{C}P^1$

(17) $$-\omega_1^3 + i\omega_2^3 = T\varphi,$$

(18) $$-\omega_1^4 + i\omega_2^4 = iT\varphi,$$

(19) $$\omega_1^\lambda = \omega_2^\lambda = 0, \ \lambda \geq 5,$$

where T is complex–valued with $|T| = \tau$. Moreover, away from the zeros of τ we can have $T = \tau$.

Upon exterior differentiation (19) yields

$$(\text{ReT } \omega_3^\lambda - \text{ImT } \omega_4^\lambda) \wedge \omega^1 - (\text{ImT } \omega_3^\lambda + \text{ReT } \omega_4^\lambda) \wedge \omega^2 = 0,$$

$$(\text{ImT } \omega_3^\lambda + \text{ReT } \omega_4^\lambda) \wedge \omega^1 + (\text{ReT } \omega_3^\lambda - \text{ImT } \omega_4^\lambda) \wedge \omega^2 = 0.$$

By Cartan's Lemma

(20) $$\text{ReT } \omega_3^\lambda - \text{ImT } \omega_4^\lambda = k_{1j}^\lambda \omega^j, \quad \text{ImT } \omega_3^\lambda + \text{ReT } \omega_4^\lambda = k_{2j}^\lambda \omega^j,$$

where $\lambda \geq 5$ and $k_{12}^\lambda + k_{21}^\lambda = 0$, $k_{11}^\lambda = k_{22}^\lambda$.

Define $L: U \to \mathbb{C}^{n-4}$ by

$$L^\lambda = k^\lambda_{12} + ik^\lambda_{11}.$$

Lemma A'. $\Lambda_2 = {}^tLL\varphi^6$ is a globally defined symmetric type (6,0) form on M.

Proof. Let e, ẽ be two Darboux frames along f as in Theorem 1. It is not hard to see that e and ẽ are related by $\tilde{e} = e\cdot h$, where

$$h = (A, B, C): U \cap \tilde{U} \to SO(2)^2 \times SO(n-4) \subset SO(n),$$

$$A = \begin{pmatrix} \cos t, -\sin t \\ \sin t, \cos t \end{pmatrix}, \; B = \begin{pmatrix} \cos s, -\sin s \\ \sin s, \cos s \end{pmatrix}.$$

Straightforward calculations give

$$\tilde{\omega}^\lambda_3 + i\tilde{\omega}^\lambda_4 = e^{-is}(C^{-1})^\lambda_\mu(\omega^\mu_3 + i\omega^\mu_4), \quad \lambda,\mu \geq 5,$$

$$\tilde{\varphi} = e^{-it}\varphi,$$

$$e^{-it}(\mathrm{Re}T + i\,\mathrm{Im}T)\tilde{L} = e^{-is}(\mathrm{Re}\tilde{T} + i\,\mathrm{Im}\tilde{T})C^{-1}L,$$

$$\mathrm{Re}\tilde{T} + i\,\mathrm{Im}\tilde{T} = e^{i(2t+s)}(\mathrm{Re}T + i\,\mathrm{Im}T).$$

It follows that $\tilde{L} = e^{2it}C^{-1}L$. Thus $\tilde{\Lambda}_2 = \Lambda_2$. □

As in Lemma B we have

Lemma B'. Λ_2 is holomorphic.

Proof. This is similar to the proof of Lemma B already given. □

Thus for $M = \mathbb{C}P^1$ we must have $\Lambda_2 \equiv 0$. Put

$$t_2 = |L|^2, \; Z_2 = \text{the zero set of } t_2.$$

Lemma C'. t_2 is of analytic type.

Proof. Omitted. □

Observation. i) $t_2 \equiv 0$ if and only if $f(M) \subset S^4 \subset S^n$.

ii) If $f: M = \mathbb{C}P^1 \to S^5$ then $t_2 \equiv 0$.

Hereafter we assume that t_2 is not identically zero. We leave the proof of the following proposition as an exercise.

Proposition. Near any point of M we can choose a Darboux frame e such that relative to e not only do we have (17–19) but also

$$iT(\omega_3^5 + i\omega_4^5) = X_2\varphi, \tag{20}$$

$$T(\omega_3^6 + i\omega_4^6) = X_2\varphi, \tag{21}$$

$$\omega_3^\lambda = \omega_4^\lambda = 0, \quad \lambda \geq 7, \tag{22}$$

where X_2 is $\mathbb{C}$–valued with $|X_2| = t_2$.

Notation. On $M\backslash Z$ put $\tau_2 = t_2/\tau$.

Inductively proceeding we may define quantities t_3, Z_3, τ_3 and so on. In summary we have

Theorem 2. Let f: $M = \mathbb{C}P^1 \to S^{2m}$ be a conformal minimal immersion which is linearly full, i.e., f(M) does not lie in a lower dimensional sphere. Then there exists a $SO(2)^m$–reduction of the SO(2m)–principal bundle $f^{-1}SO(2m+1) \to M$, denoted by $\mathscr{F} \to M$, such that relative to any local section e of $\mathscr{F} \to M$ we have

$$\omega_0^a = 0, \ a > 2; \tag{23}$$

$$\omega_{2i+1}^{\lambda_i} = \omega_{2i+2}^{\lambda_i} = 0, \ 1 \leq i \leq m-3, \ \lambda_i > 2i+4; \tag{24}$$

$$-\omega_1^3 + i\omega_2^3 = T\varphi, \ -\omega_1^4 + i\omega_2^4 = iT\varphi,$$

$$iT(\omega_3^5 + i\omega_4^5) = X_2\varphi, \ T(\omega_3^6 + i\omega_4^6) = X_2\omega, \ \cdots. \tag{25}$$

Moreover, near any point in $M\backslash\{Z \cup Z_2 \cup \cdots \cup Z_{m-1}\}$ we can further choose e to achieve

$$(26)\qquad \begin{aligned} -\omega_1^3 + i\omega_2^3 &= \tau\varphi, \quad -\omega_1^4 + i\omega_2^4 = i\tau\varphi, \\ \omega_3^5 + i\omega_4^5 &= -i\tau_2\varphi, \quad \omega_3^6 + i\omega_4^6 = \tau_2\varphi, \\ &\vdots \\ \omega_{2m-3}^{2m-1} + i\omega_{2m-2}^{2m-1} &= -i\tau_{m-1}\varphi, \quad \omega_{2m-1}^{2m} + i\omega_{2m-2}^{2m} = \tau_{m-1}\varphi. \end{aligned}$$

Remark. i) We do not lose any generality by assuming that $n = 2m$ since a minimal two-sphere must lie in an even dimensional sphere as indicated earlier.

ii) τ_i is defined on $M\backslash\{Z \cup Z_1 \cup \cdots \cup Z_{i-1}\}$ and the zero set of τ_i is Z_i. However, we will see shortly that (using the fact that the Z_i's are all discrete) each τ_i^2 is, in fact, a smooth function on M (i.e., it can be uniquely extended to such a function) so that τ_i is continuous everywhere and smooth away from its zeros.

iii) As a consequence of Theorem 2 we have, given a conformal minimal immersion $f: \mathbb{C}P^1 \to S^n$, a global lifting $\Phi_f: \mathbb{C}P^1 \to SO(2m+1)/SO(2)^m$. In fact, it can be shown that with respect to the natural complex structure of $SO(2m+1)/SO(2)^m$ the map Φ_f is holomorphic.

iv) We also get well-defined maps

$$f_i: M \to G_{2m+1,2} = Q_{2m-1} \subset \mathbb{C}P^{2m}, \quad 1 \leq i \leq m,$$

given by $f_i = [e_{2i-1} \wedge e_{2i}]$, where $e = (e_\alpha)$ is any local section of $\mathcal{F} \to M$. The conjugate of f_m is usually called the directrix curve of f. It is routinely verified that $\bar{f}_m$ is holomorphic. We will also see in Chapter V that f_i, $1 \leq i \leq m-1$, are all conformal minimal immersions possibly with isolated branch points.

v) Theorem 2 remains valid even without the assumption $M = \mathbb{C}P^1$ given that the holomorphic symmetric differentials $\Lambda, \Lambda_1, \cdots, \Lambda_{m-1}$ all vanish on M. A conformal minimal immersion for which this happens is also called a superminimal immersion (Bryant) or a pseudoholomorphic immersion (Calabi).

The SO(2m)–principal bundle $\mathscr{F} \to M = \mathbb{C}P^1$ gives rise to a Whitney sum decomposition of the normal bundle $TM^{\perp} \to M$:

(27) $$TM^{\perp} = L_1 \oplus \cdots \oplus L_{m-1},$$

where $L_i = \text{span}\,\{e_{2i+1}, e_{2i+2}\}$, and e, a local section of $\mathscr{F} \to M$. The induced connection ∇_i on $L_i \to M$ is given by

(28) $$\nabla_i e_{2i+1} = e_{2i+2} \otimes \omega^{2i+2}_{2i+1},\ \nabla_i e_{2i+2} = e_{2i+1} \otimes \omega^{2i+1}_{2i+2} \text{ (no sum)}.$$

Recall that the Levi–Civita connection of (S^{2m}, ds^2_E) is given by

$$\nabla e_\alpha = e_\beta \otimes \omega^\beta_\alpha,\quad 0 \le \alpha,\beta \le 2m.$$

Notation. K_i = the Gaussian curvature of (L_i, ∇_i).

Lemma. Maintaining the notations in Theorem 2 we have

(29) $$K_i = 2(\tau^2_{i+1} - \tau^2_i),\ 1 \le i \le m-1,\ \tau_m = 0.$$

Proof. Have $d\omega^{2i+1}_{2i+2} = K_i\, \omega^1 \wedge \omega^2$ since the induced connection form of $L_i \to M$ with respect to the framing $\{e_{2i+1}, e_{2i+2}\}$ is $e^*\Omega^{2i+1}_{2i+2} = \omega^{2i+1}_{2i+2}$. On the other hand using the Maurer–Cartan structure equations of SO(2m+1) combined with the relations in (25–26) we get $d\omega^{2i+1}_{2i+2} = 2(\tau^2_{i+1} - \tau^2_i)\, \omega^1 \wedge \omega^2$. It follows that (29) holds in $M\backslash\{Z_1 \cup \cdots \cup Z_{m-1}\}$. But the K_i's are continuous in all of M and hence (29) must hold in all of M also by continuity. □

The above proof shows that the τ_i's are smooth functions on $M = \mathbb{C}P^1$.

Proposition. Maintaining the notations of Theorem 2 we have

(30) $$\Delta \log \tau_i = K + K_i - K_{i-1},$$

where $1 \le i \le m-1$, $\tau_1 = \tau$, and $K_0 = -K$.

Proof. From (26) we have

(*) $$-\omega_1^3 = \tau\omega^1, \quad \omega_2^3 = \tau\omega^2,$$

(**) $$\omega_1^4 = \tau\omega^2, \quad \omega_2^4 = \tau\omega^1.$$

Upon exterior differentiation (*) yields

(a) $$-(d \log\tau) \wedge \omega^1 + (2\omega_2^1 + \omega_4^3) \wedge \omega^2 = 0.$$

Likewise (**) gives

(b) $$d \log\tau \wedge \omega^2 + (2\omega_2^1 + \omega_4^3) \wedge \omega^1 = 0.$$

Putting (a) and (b) together,

$$[d \log\tau + i(2\omega_2^1 + \omega_4^3)] \wedge \varphi = 0.$$

It follows that

$$2\omega_2^1 + \omega_4^3 = {*d} \log\tau,$$

where * is the Hodge star operator of (M, ds^2). Now

$$d{*d} \log\tau = \Delta \log\tau \, \omega^1 \wedge \omega^2,$$

and we thus obtain $\Delta \log\tau = 2K + K_1$ establishing (30) for $i = 1$. The rest is similar. We record the formula

$$*d \log\tau_i = \omega_2^1 - \omega_{2i}^{2i-1} + \omega_{2i+2}^{2i+1}, \quad i \geq 2. \quad \square$$

From (30) we obtain

$$(m-1)\Delta\log\tau_1 + (m-2)\Delta\log\tau_2 + \cdots + \Delta\log\tau_{m-1} =$$

$$\tfrac{1}{2}(m-1)(m+2)K + K_1 + \cdots + K_{m-1}.$$

But $K_1 + K_2 + \cdots + K_{m-1} = -2\tau^2 = K - 1$, hence giving

(31) $$\Delta \log(\tau_1^{m-1}\tau_2^{m-2}\cdots\tau_{m-1}) = \tfrac{1}{2}\, m(m+1)K - 1.$$

Corollary. Let $f\colon \mathbb{C}P^1 = M \to S^{2m}$ be a linearly full conformal minimal immersion. We then have

i) $A(f) \geq 2\pi m(m+1)$ ($A(f)$ = the area of f);

ii) if $K \geq \frac{2}{m(m+1)}$ then $K \equiv \frac{2}{m(m+1)}$;

iii) if $A(f) = 2\pi m(m+1)$ and $K \leq \frac{2}{m(m+1)}$ then $K \equiv \frac{2}{m(m+1)}$.

Proof. Rewriting (31) we get

(*) $\frac{1}{2}m(m+1)K = 1 + (m-1)\Delta\log\tau_1 + (m-2)\Delta\log\tau_2 + \cdots + \Delta\log\tau_{m-1}$.

Now the Euler characteristic of $M = \mathbb{C}P^1 = 2$ and the Gauss–Bonnet theorem states that

$$2 = \frac{1}{2\pi}\int_M K\,\omega^1\wedge\omega^2.$$

Also

$$A(f) = \int_M \omega^1\wedge\omega^2.$$

We put

$$\#_i = \frac{-1}{2\pi}\int_M \Delta\log\tau_i\,\omega^1\wedge\omega^2,$$

where $1 \leq i \leq m-1$ and $\#_0 = \#_m = 0$. $\#_i$ is the total order of the zeros of τ_i by the elementary argument principle. In particular, $\#_i \geq 0$. Integration of (*) now yields

$$2\pi m(m+1) = A(f) - 2\pi\cdot[(m-1)\#_1 + (m-2)\#_2 + \cdots + \#_{m-1}].$$

So $A(f) = 2\pi m(m+1) + c$ with $c \geq 0$ proving (i). To prove (ii) put

$$h = \log(\tau_1^{m-1}\tau_2^{m-2}\cdots\tau_{m-1}).$$

By hypothesis and by (31) $\Delta h \geq 0$. So h is a subharmonic function with singularities at $Z_1 \cup \cdots \cup Z_{m-1}$ where it goes to $-\infty$. Thus h attains a maximum in M and the maximum principle for subharmonic functions implies that h must be constant. (ii) now follows easily. In (iii) note that h is superharmonic. Now $A(f) = 2\pi m(m+1)$ and this forces $\#_i = 0$ for every i. So

$Z_i = \emptyset$ for every i and h attains a minimum on M. The minimum principle for superharmonic functions does the rest. □

Note that Corollary (ii) remains valid for superminimal immersions in general as we did not have to use the assumption $M = \mathbb{C}P^1$ directly.

Exercise. Let f, $\tilde{f}: \mathbb{C}P^1 \to S^{2m}$ be two *isometric* linearly full conformal minimal immersions. Then they are congruent to each other, i.e., there exists a fixed $A \in SO(2m+1)$ with $\tilde{f} = A \cdot f$.

§3. The Twistor Fibration

The quaternion algebra $\mathbb{H}$ is defined as a 4–dimensional real vector space with the basis 1, i, j, k, and with the ring multiplication coming from

$$i^2 = j^2 = k^2 = -1,\ ij = k = -ji,\ jk = i = -kj,\ ki = j = -ik.$$

$\mathbb{H}$ is also a two–dimensional right module over $\mathbb{C}$ with the basis 1, j. Have

$$xy = (z_1w_1 - \bar{z}_2w_2) + j(z_2w_1 + \bar{z}_1w_2),$$
$$\text{for } x = z_1 + jz_2,\ y = w_1 + jw_2 \in \mathbb{H}.$$

$\mathbb{H}$ can also be made into a left $\mathbb{C}$–vector space by the rule

$$zj = j\bar{z}, \quad z \in \mathbb{C}.$$

There is a natural conjugation in $\mathbb{H}$, denoted by *,

$$(z_1 + jz_2)^* = \bar{z}_1 - jz_2.$$

Then $(xy)^* = y^*x^*$.

Convention. We will always think of $\mathbb{H}$ as a right $\mathbb{C}$–module and use the identification $\mathbb{H} = \mathbb{C}^2$ given by $(z_1 + jz_2) \leftrightarrow {}^t(z_1, z_2)$.

An element of $\mathbb{H}$ is called a quaternion. $\mathbb{H}^n$ denotes the set of n–tuples of quaternions written as column vectors. We will think of $\mathbb{H}^n$ as a right $\mathbb{H}$–module as well as a right $\mathbb{C}$–module. Identify $\mathbb{H}^n$ with $\mathbb{C}^{2n}$ via

$$z_1 + jz_2 \mapsto {}^t(z_1, z_2),$$

where now $z_i \in \mathbb{C}^n$. Define a $\mathbb{H}$–valued bilinear form on $\mathbb{H}^n$ by

$$(\ ,\ ): \mathbb{H}^n \times \mathbb{H}^n \to \mathbb{H}, \quad (x, y) = {}^t x^* y.$$

The real part of (,) is the standard inner product on $\mathbb{R}^{4n} \cong \mathbb{H}^n$,

$$\mathrm{Re}(x,y) = \tfrac{1}{2}\,[(x,y) + (x,y)^*].$$

The "isometry" group of $(\mathbb{H}^n, (,))$ is Sp(n). So

(1) $$\mathrm{Sp}(n) = \{A \in \mathrm{GL}(n;\mathbb{H}): {}^t A^* A = I\}.$$

Remark. i) There are the well–known isomorphisms $\mathrm{Sp}(1) \cong \mathrm{Spin}(3)$ and $\mathrm{Sp}(2) \cong \mathrm{Spin}(5)$, where Spin(n) is the simply connected double cover of $\mathrm{SO}(n;\mathbb{R})$.

ii) Also $\mathrm{Sp}(1) = \{x \in \mathbb{H}: x^* x = 1\} \cong S^3 \cong \{\begin{pmatrix} z, & w \\ -\bar{w}, & \bar{z}\end{pmatrix}: z,w \in \mathbb{C},\ |z|^2 + |w|^2 = 1\}$ $= \mathrm{SU}(2)$. Sp(1) is the set of unit quaternions.

iii) There are the inclusions $U(n) \subset \mathrm{Sp}(n)$ coming from

$$\mathbb{C} \hookrightarrow \mathbb{H},\ z \mapsto z + j\cdot 0.$$

iv) One has the monomorphism coming from $\mathbb{H}^n = \mathbb{C}^{2n}$

(2) $$\mathrm{Sp}(n) \hookrightarrow \mathrm{SU}(2n) \text{ via } X + jY \mapsto \begin{pmatrix} X, & -\bar{Y} \\ Y, & \bar{X}\end{pmatrix}.$$

Quaternionic projective space $\mathbb{HP}^n$ is the space of $\mathbb{H}$–lines in $\mathbb{H}^{n+1}$. There is the projection

(3) $$\mathbb{H}^{n+1}\backslash\{0\} \to \mathbb{HP}^n, \quad v \mapsto [v]_H.$$

$\mathbb{HP}^n$ is simply connected and as a homogeneous space

(4) $$\mathbb{HP}^n = \mathrm{Sp}(n+1)/\mathrm{Sp}(1)\times\mathrm{Sp}(n).$$

More precisely, for $A \in \mathrm{Sp}(n+1)$, $[v]_H \in \mathbb{HP}^n$,

(5) $$A\cdot[v]_H = [Av]_H,$$

and under this action $Sp(1)\times Sp(n)$ is the isotropy subgroup at $0 = [{}^t(1,0,\cdots,0)]_H$, where $Sp(1)\times Sp(n) \subset Sp(n+1)$ via $(a,A) \mapsto \begin{pmatrix} a,0 \\ 0,A \end{pmatrix}$.

Definition. The twistor map is a fibration

(6) $$\pi\colon \mathbb{C}P^{2n+1} \to \mathbb{H}P^n,$$

$$[{}^t(z_1, z_2)] \mapsto [z_1 + jz_2]_H, \quad z_1, z_2 \in \mathbb{C}^{n+1}.$$

The fibre of π at a point $[v]_H$ is just the set of complex lines in $[v]_H \cong \mathbb{H} \cong \mathbb{C}^2$, hence is $\mathbb{C}P^1$. Keeping in mind the injection $Sp(n+1) \subset SU(2n+2)$ we have the following commutative diagram of projections.

$$\begin{array}{ccc} & Sp(n+1) & \\ {}^{\pi_2}\swarrow & & \searrow^{\pi_1} \\ \mathbb{H}P^n & \xleftarrow[\pi]{} & \mathbb{C}P^{2n+1} \end{array}$$

$$\pi_1(A) = \begin{pmatrix} X, -\bar{Y} \\ Y, \bar{X} \end{pmatrix} {}^t[1,0,\cdots,0] = \begin{bmatrix} X_0 \\ Y_0 \end{bmatrix},$$

where $A = X + jY$ and $X = (X_0,\cdots, X_n)$, $Y = (Y_0,\cdots,Y_n)$;

$$\pi_2(A) = A\cdot {}^t[1,0,\cdots,0]_H = [A_0]_H, \quad A = (A_0,\cdots,A_n).$$

It follows that $\pi\circ\pi_1(A) = [X_0 + jY_0]_H = [A_0]_H$. Note that $Sp(n+1) \to \mathbb{C}P^{2n+1}$ is a principal $U(1)\times Sp(n)$-bundle.

Definition. The horizontal distribution $\mathcal{H}$ on $\mathbb{C}P^{2n+1}$ is defined to be the orthogonal (with respect to the Fubini–Study metric) complement to the fibre of $\pi\colon \mathbb{C}P^{2n+1} \to \mathbb{H}P^n$.

Take $[v]_H \in \mathbb{H}P^n$, $v = z_1 + jz_2$. Then

(7) $$\pi^{-1}([v]_H) = \{a\,{}^t[z_1,z_2] + b\,{}^t[-\bar{z}_2,\bar{z}_1]\colon a,b \in \mathbb{C}\}.$$

(Note that $(z_1 + jz_2)j = -\bar{z}_2 + j\bar{z}_1$.) So $\pi^{-1}([v]_H)$ is the line in $\mathbb{C}P^{2n+1}$ through ${}^t[z_1,z_2]$ and ${}^t[-\bar{z}_2,\bar{z}_1]$.

We have a $\mathbb{C}$–isomorphism

(8) $$T^{1,0}_{[v]}\,\mathbb{C}P^{2n+1} \to v^{\perp} \subset \mathbb{C}^{2n+2}, \quad v \in \mathbb{C}^{2n+2}\backslash\{0\},$$

where $v^{\perp}$ denote the orthogonal complement to v in $\mathbb{C}^{2n+2}$. Under this isomorphism the Fubini–Study metric on $\mathbb{C}P^{2n+1}$ becomes the standard hermitian inner product on $\mathbb{C}^{2n+2}$ restricted to the $v^{\perp}$'s. Write $v = {}^t(z_1, z_2) \in \mathbb{C}^{2n+2}$ and consider the holomorphic 1–form σ on $\mathbb{C}^{2n+2}$ given by

$$\sigma_v = -{}^tz_2dz_1 + {}^tz_1dz_2 = \langle {}^t(-\bar{z}_2,\bar{z}_1), dv\rangle,$$

where $\langle,\rangle$ is the standard hermitian inner product in $\mathbb{C}^{2n+2}$. Then under the identification given by (8) we have

(9) $$\mathcal{H}_{[v]} \leftrightarrow \{w \in v^{\perp}: \sigma_v(w) = 0\}.$$

Let U_0 denote the affine part of $\mathbb{C}P^{2n+1}$ about $0 = {}^t[1,0,\cdots,0]$ and write

$$\mathbb{C}P^{2n+1} = U_0 \amalg \mathbb{C}P^{2n}_{\infty}.$$

The corresponding inhomogeneous coordinates are (z_1^a, z_2^0, z_2^a), $1 \leq a \leq n$, $z_1^0 = 1$.

Notation. $z_1^a = t^a$, $z_2^a = s^a$, $z_2^0 = s^0$.

On $U_0 \cong \mathbb{C}^{2n+1}$, $\mathcal{H}$ is given by a single exterior equation

(10) $$\mathcal{H}_0 = \{\sigma_0 = ds^0 + \Sigma\, t^a ds^a - \Sigma\, s^a dt^a = 0\}.$$

Definition. Let M be a (connected as always) Riemann surface. A holomorphic map $h: M \to \mathbb{C}P^{2n+1}$ tangential to $\mathcal{H}$ will be called a horizontal curve.

Remark. i) $\mathcal{H}$ is not completely integrable. In fact $\sigma_0 \wedge (d\sigma_0)^n \neq 0$ and σ_0 is a contact form on U_0.

ii) If a projective line ($\cong \mathbb{C}P^1$) in $\mathbb{C}P^{2n+1}$ is orthogonal to the fibre of π: $\mathbb{C}P^{2n+1}$

$\to \mathbb{H}P^n$ at a single point then it is orthogonal to the fibre of π everywhere. Thus such a line is a horizontal curve.

The following result of Bryant gives a useful characterization of horizontal curves in $\mathbb{C}P^3$.

Proposition. Let f and g be meromorphic functions on M with g nonconstant. Define a holomorphic map $\Phi(f, g)$: $M \to \mathbb{C}P^3$ by

$$\Phi(f, g) = {}^t[1, g, f - \tfrac{1}{2}\cdot g\cdot \tfrac{df}{dg}, \tfrac{1}{2}\cdot\tfrac{df}{dg}]. \tag{11}$$

Then Φ is a horizontal curve. Conversely, any nonconstant hotizontal curve $M \to \mathbb{C}P^3$ arises in this manner for some unique meromorphic functions f and g on M or else it is contained in a line in $\mathbb{C}P^3$.

Proof. On the affine part U_0 we have

$$\sigma_0 = ds^0 + t^1 ds^1 - s^1 dt^1.$$

It is a routine computation to check that $\Phi(f,g)^* \sigma_0 = 0$ showing that $\Phi(f,g)$ is horizontal on $M\backslash\Phi^{-1}(\mathbb{C}P^2_\infty)$. But the distribution $\mathscr{H}$ is smooth and we must have $\Phi(f,g)^* \sigma = 0$ everywhere. Conversely, suppose we are given a horizontal curve Φ: $M \to \mathbb{C}P^3$. First suppose that $\Phi(M) \cap U_0 = \emptyset$. Write in U_0, $\Phi = {}^t[1, t^1, s^0, s^1]$. If $t^1 = a =$ constant then $0 = \Phi^* \sigma_0 = d(s^0 + as^1)$. Thus $s^0 + as^1 =$ constant $=$ b. But this says $\Phi(M) \subset \{{}^t[1, a, b-as, s]: s \in \mathbb{C}\}$, a line in $\mathbb{C}$. We now look at the case where t^1 is not constant. Put $f = s^0 + t^1 s^1$, $g = t^1$. Then $0 = \Phi^* \sigma_0$ $= df - 2s^1 dg$. So $s^1 = \frac{1}{2}\cdot\frac{df}{dg}$ and $s^0 = f - \frac{1}{2}\cdot g\cdot\frac{df}{dg}$ and $\Phi = \Phi(f,g)$. It remains to consider the case $\Phi(M) \subset \mathbb{C}P^2_\infty = \{{}^t[z_1^0, z_1^1, z_2^0, z_2^1]: z_1^0 = 0\}$. Without loss of generality we further suppose that $\Phi(M)$ is not contained in the line $\{z_1^0 = z_1^1 = 0\}$. Put $U_1 = \{z_1^1 \neq 0\}$ and also let $T^0 = z_1^0/z_1^1$, $S^0 = z_2^0/z_1^1$, $S^1 = z_2^1/z_1^1$ be the corresponding inhomogeneous coordinates. Write Φ in U_1 as $\Phi = {}^t[0, 1, S^0, S^1]$, with the usual abuse of notation. Then in U_1, $0 = \Phi^*\sigma = dS^1$. Thus the meromorphic function S^1, i.e., $S^1 \circ \Phi$ is constant, say c. It follows that $\Phi(M)$ is contained in the line $\{z_1^0 = 0, z_1^1 = 1, z_2^1 = c\}$. □

A pair of meromorphic functions f, g on a compact Riemann surface M corresponds to a holomorphic map Ψ: $M \to \mathbb{C}P^2$ by $\Psi = {}^t[g,f,1]$ (using again the fact that a rational map from a Riemann surface is holomorphic) and the above proposition gives us a correspondence between plane algebraic curves and horizontal space curves.

Theorem (Bryant). Let M be a *compact* Riemann surface. Then there exists a holomorphic horizontal embedding $M \hookrightarrow \mathbb{C}P^3$.

Proof. As an application of the Riemann–Roch theorem we can holomorphically map $M \to \mathbb{C}P^2$ such that only singular points are ordinary double points. Fix such a holomorphic map Ψ: $M \to \mathbb{C}P^2$ and put $C = \Psi(M)$. Choose $q_0 \in \mathbb{C}P^2$ such that

a) $q_0 \notin C$;

b) q_0 does not lie in any flex or bitangent to C;

c) q_0 does not lie in the tangent cone to any double point of C.

Also let L be a line through q_0 which is not tangent to C nor does it pass through a double point of C. Let q_1 be another point on L not on C. Let $\{p_1,\cdots,p_a\}$ be the set of points of M mapped to L. Also let $\{q_1,\cdots,q_b\}$ be the set of points of M such that the tangent to C at $\Psi(q_j)$ passes through q_0. Finally choose homogeneous coordinates on $\mathbb{C}P^2$ such that $q_0 = {}^t[0,1,0]$, $q_1 = {}^t[1,0,0]$ and write $\Psi = {}^t[g,f,1]$ for unique meromorphic functions on M.

Index Notation. $1 \le i \le a$, $1 \le j \le b$.

We then have the following.

i) If $p \notin \{p_i, q_j\}$ then f and g are holomorphic at p and $dg(p) \neq 0$;

ii) Both f and g have a simple pole at p_i. In fact if we let $z(p) = 1/g(p)$ for p near p_i then z is a holomorphic coordinate about p_i. So there exists a holomorphic function $F_i(z)$ near $0 \leftrightarrow p_i$ with

$$f(p) = F_i(z(p))/z(p).$$

Since $q_0, q_1 \notin \{p_i\}$ and $L \cap C$ transversally the values $\{F_i(0)\}$ are nonzero, finite and distinct;

iii) Near q_j there exists a holomorphic coordinate $z(p)$ unique upto sign so that

$g(p) = A_j + 0 \cdot z(p) + \text{higher order terms}$,

$f(p) = \tilde{F}_j(z(p))$

for unique constants A_j and local holomorphic functions $\tilde{F}_j$. (q_0 does not lie on a flex tangent to C whch implies that dg vanishes only to first order.) Since Ψ is immersive away from the singular set and q_0 does not lie on a tangent cone to a double point we must have $\tilde{F}_j'(0) \neq 0$. Since L is not tangent to C and q_0 lies on no bitangent the A_j's are finite and distinct. We now *claim* that $\Phi(f,g)\colon M \to \mathbb{C}P^3$ is a holomorphic embedding. We look at three cases.

1) Suppose $\Phi(f,g) \in U_0$ (the affine part of $\mathbb{C}P^3$ about ${}^t[1,0,0,0]$). This is so iff $p \notin \{p_i, q_j\}$. Note that $d\Phi \neq 0$ at p since $dg \neq 0$ at p. Suppose we have $\Phi(p) = \Phi(q)$ for $p,q \in M\backslash\{p_i, q_j\}$. Then we must have

$$g(p) = g(q),\ \ f(p) = f(q),\ \ \frac{df}{dg}(p) = \frac{df}{dg}(q).$$

So $\Phi(p) = \Phi(q)$ and the tangents to C at p and q are the same. But only singular points of C are ordinary double points, hence $p = q$.

2) Let $z = z(p) = 1/g(p)$, a holomorphic coordinate about p_i. Then

$$\Phi(p) = {}^t[\tfrac{1}{g}(1,\ g,\ f - \tfrac{1}{2}\cdot g\cdot \tfrac{df}{dg},\ \tfrac{1}{2}\cdot\tfrac{df}{dg})]$$

$$= {}^t[z,\ 1,\ F_i(z) - \tfrac{1}{2}(F_i(z) - F_i'(z)\cdot z),\ \tfrac{1}{2}\cdot z(F_i(z) - F_i'(z)z)].$$

So $d\Phi \neq 0$ at p_i and

$$\Phi(p_i) = {}^t[0,\ 1,\ F_i(0)/2,\ 0].$$

3) If $z = z(p)$ is a holomorphic coordinate about q_j as in iii) we then have

$$\Phi(p) = {}^t[z,\ zg,\ z\tilde{F}_j(z) - \tfrac{1}{2}\cdot g\tilde{F}_j'(z),\ \tilde{F}_j'(z)/2],\ p \text{ near } q_j.$$

So, $\Phi(q_j) = {}^t[0,\ 0,\ -A_j = -g(0),\ 1]$ since $\tilde{F}_j'(0) \neq 0$.

We thus see that the points p_i, q_j are sent to distinct points in $\mathbb{C}P^2_\infty$ proving that Φ is one to one (remember that the singuar set $\subset \{p_i, q_j\}$). This finishes the proof as we have also shown that Φ is an immersion. □

§4. Minimal Surfaces in $\mathbb{H}P^1$

Let Ω denote the Maurer–Cartan form of Sp(n+1). Ω is a $\mathfrak{sp}$(n+1)–valued left–invariant 1–form on Sp(n+1) given by

$$\Omega_g(X) = L_{g^{-1}*}X,\quad X \in T_g Sp(n+1).$$

Write $\Omega = (\Omega^\alpha_\beta) = g^{-1}dg$ using the matrix notation. Then $\Omega^\alpha_\beta = -\Omega^{\beta *}_\alpha$, where $*$ is the quaternionic conjugation. Put

(1) $$\Omega^\alpha_\beta = \Gamma^\alpha_\beta + j\Sigma^\alpha_\beta,$$

for complex–valued 1–forms Γ^α_β and Σ^α_β. Then

$$\bar{\Gamma}^\alpha_\beta = -\Gamma^\beta_\alpha,\quad \Sigma^\alpha_\beta = \Sigma^\beta_\alpha,$$

i.e., Γ is $\mathfrak{u}$(n+1)–valued and Σ is $\mathcal{S}$(n+1;$\mathbb{C}$)–valued. The Maurer–Cartan structure equations become

(2) $$d\Gamma = -\Gamma \wedge \Gamma + \bar{\Sigma} \wedge \Sigma, \quad d\Sigma = -\bar{\Gamma} \wedge \Sigma - \Sigma \wedge \Gamma.$$

Recall from §3 that $G_0 = Sp(1)\times Sp(n) \subset Sp(n+1)$ $((a,A) \leftrightarrow (\begin{smallmatrix} a,0 \\ 0,A \end{smallmatrix}))$, where $0 = {}^t[1,0,\cdots,0]_H \in \mathbb{H}P^n$. There is the vector space direct sum decomposition $\mathfrak{g} = \mathfrak{g}_0 \oplus \mathfrak{m}$, where

(3) $$\mathfrak{m} = \left\{ \begin{bmatrix} 0, -{}^tX^* \\ X, \ 0 \end{bmatrix} : X \in \mathbb{H}^n \right\}.$$

We will identify $\mathfrak{m}$ with $\mathbb{H}^n$ via $[\cdot] \leftrightarrow X$. The Adjoint action of G_0 on $\mathfrak{m}$ is given by the formula

(4) $$\mathrm{Ad}(k)X = AXa^{-1},$$

where $k = (a, A) \in Sp(1)\times Sp(n)$. From this formula we can see that the inner product on $\mathfrak{m}$ given by

(5) $$\langle\ ,\ \rangle = \sum_1^n \Omega_0^{a*} \cdot \Omega_0^a \quad \text{(symmetric product)}$$

is $\mathrm{Ag}(G_0)$–invariant and hence it gives an invariant metric on $\mathbb{H}P^n$. (Since $\mathbb{H}P^n$ is a Riemannian symmetric space any other invariant metric is a constant multiple of the one obtained above.)

Hereafter we deal exclusively with the case $n = 1$. Take a local section s of $Sp(2) \to \mathbb{H}P^1$.

Notation. $s^*\Gamma = \gamma$, $s^*\Sigma = \sigma$.

On $\mathbb{H}P^1$ we use

(6) $$ds^2 = 4(\gamma_0^1 \cdot \bar{\gamma}_0^1 + \sigma_0^1 \cdot \bar{\sigma}_0^1).$$

We will explain the presence of "4" shortly.

Notation. $\varphi^1 = 2\mathrm{Re}\ \gamma_0^1$, $\varphi^2 = 2\mathrm{Im}\ \gamma_0^1$, $\varphi^3 = 2\mathrm{Re}\ \sigma_0^1$, $\varphi^4 = 2\mathrm{Im}\ \sigma_0^1$.

We now compute the Levi–Civita connection forms of $(\mathbb{H}P^1, ds^2)$ relative to the coframe $(\varphi^1, \cdots, \varphi^4)$. Using (2) we obtain

$$d\gamma_0^1 = (\gamma_0^0 - \gamma_1^1) \wedge \gamma_0^1 - \sigma_0^0 \wedge \bar{\sigma}_0^1 + \bar{\sigma}_1^1 \wedge \sigma_0^1,$$

$$d\sigma_0^1 = (\gamma_0^0 + \gamma_1^1) \wedge \sigma_0^1 - \sigma_0^0 \wedge \bar{\gamma}_0^1 + \sigma_1^1 \wedge \gamma_0^1.$$

Taking the real and imaginary parts we get

$$d\varphi^\alpha = -\omega_\beta^\alpha \wedge \varphi^\beta \ (1 \le \alpha,\beta \le 4), \tag{7}$$

where

$$\omega_2^1 = -i(\gamma_0^0 - \gamma_1^1),\ \ \omega_3^1 = \mathrm{Re}\ \sigma_0^0 - \mathrm{Re}\ \sigma_1^1,\ \ \omega_4^1 = -\mathrm{Im}\ \sigma_1^1 + \mathrm{Im}\ \sigma_0^0,$$

$$\omega_3^2 = \mathrm{Im}\ \sigma_0^0 + \mathrm{Im}\ \sigma_1^1,\ \ \omega_4^2 = -(\mathrm{Re}\ \sigma_0^0 + \mathrm{Re}\ \sigma_1^1),\ \ \omega_4^3 = -i(\gamma_0^0 + \gamma_1^1).$$

From this we further calculate that

$$d\omega_\beta^\alpha = -\omega_\gamma^\alpha \wedge \omega_\beta^\gamma + \varphi^\alpha \wedge \varphi^\beta, \tag{8}$$

showing that $(\mathbb{H}P^1, ds^2)$ is of constant sectional curvature 1. Now both $\mathbb{H}P^1$ and S^4 are compact and simply connected. Therefore $(\mathbb{H}P^1, ds^2)$ is isometric to the standard unit sphere $S^4 \subset \mathbb{R}^5$.

Consider a conformal immersion $f: M \to \mathbb{H}P^1 \cong S^4$ from a Riemann surface M. Taking a moving frame $e: U \subset M \to Sp(2)$ along f and write, abusing notation slightly,

$$e^*\Gamma = \gamma,\ \ e^*\Sigma = \sigma. \tag{9}$$

(We will confuse $e^*\Omega$ and $s^*\Omega$ unless there is a real danger of confusion.) Also write $2\mathrm{Re}\ e^*\Gamma_0^1 = \varphi^1$, $2\mathrm{Im}\ e^*\Gamma_0^1 = \varphi^2$, $2\mathrm{Re}\ \sigma_0^1 = \varphi^3$, $2\mathrm{Im}\ \sigma_0^1 = \varphi^4$.

Lemma. There exists a moving frame $\tilde{e}$ along f about any point of M such that

$$\tilde{e}^*\Sigma_0^1 = 0. \tag{10}$$

Proof. Let $\tilde{e} = e\cdot k$ be another moving frame along f with $k: U \cap \tilde{U} \to Sp(1)\times Sp(1)$, $k = (a, A)$. Then

$$\tilde{e}^*\Omega_{\mathfrak{m}} = \mathrm{Ad}(k^{-1})e^*\Omega_{\mathfrak{m}}. \tag{11}$$

Using (4) now we see that we can choose k such that (10) holds. □

A moving frame along f: M → $\mathbb{H}P^1$ as in the above lemma will be called a *symplectic frame* alnog f. So for a symplectic frame *e* along f, $\sigma_0^1 = 0$ and the induced metric on M is given by

$$ds_M^2 = 4\gamma_0^1\cdot\bar{\gamma}_0^1 \doteq (\varphi^1)^2 + (\varphi^2)^2. \tag{12}$$

Take a symplectic frame *e* and put $\varphi = \varphi^1 + i\varphi^2$ ($\varphi^1 = 2\mathrm{Re}\, e^*\Gamma_0^1$, etc). φ is a type (1,0) form on M since f is conformal.

Consider a local coframe in ($\mathbb{H}P^1$, ds^2) given by ($\hat{\varphi}^1 = 2\mathrm{Re}\; s^*\Gamma_0^1, \cdots, \hat{\varphi}^4 = 2\mathrm{Im}\; s^*\Sigma_0^1$). Assume that $(\hat{\varphi}^\alpha)$ are so chosen that $f^*\hat{\varphi}^\alpha = \varphi^\alpha$, where we write the φ^α's relative to a symplectic frame *e* along f (use the usual argument about extending vector fields locally). In particular, $f^*\hat{\varphi}^3 = f^*\hat{\varphi}^4 = 0$. It now follows from (7) that the second fundamental forms II^3 and II^4 on M written relative to $(\hat{\varphi}^\alpha)$ are given by

$$\begin{aligned} II^3 &= (-\mathrm{Re}\;\sigma_0^0 + \mathrm{Re}\;\sigma_1^1)\varphi^1 - (\mathrm{Im}\;\sigma_1^1 + \mathrm{Im}\;\sigma_0^0)\varphi^2, \\ II^4 &= (\mathrm{Im}\;\sigma_1^1 - \mathrm{Im}\;\sigma_0^0)\varphi^1 + (\mathrm{Re}\;\sigma_0^0 + \mathrm{Re}\;\sigma_1^1)\varphi^2. \end{aligned} \tag{13}$$

Define local $\mathbb{C}$–valued functions a, b, c by

$$\sigma_1^1 = a\varphi + b\bar{\varphi}, \quad -\sigma_0^0 = b\varphi + c\bar{\varphi}, \tag{14}$$

and put

$$II = \sigma_1^1\varphi - \sigma_0^0\bar{\varphi} = II^3 + iII^4. \tag{15}$$

Lemma. f is minimal if and only if $b \equiv 0$.

Proof. $II = a\varphi\cdot\varphi + 2b\varphi\cdot\bar{\varphi} + c\bar{\varphi}\cdot\bar{\varphi} = (h_{ij}^3 + h_{ij}^4)\varphi^i\cdot\varphi^j$ ($1 \le i,j \le 2$), where $II^a = h_{ij}^a\varphi^i\cdot\varphi^j$. It follows that

$$h^3 + ih^4 = \begin{bmatrix} a + 2b + c, & i(a - c) \\ i(a - c), & -a + 2b - c \end{bmatrix}.$$

So trace h^3 + i trace h^4 = 4b and the result follows. □

Theorem. Let Φ: M → $\mathbb{C}P^3$ be a horizontal holomorphic immersion from a Riemann surface M. Then $\pi\circ\Phi$: M → $\mathbb{H}P^1$ is a conformal minimal immersion, where π: $\mathbb{C}P^3 \to \mathbb{H}P^1$ is the twistor fibration.

Proof. Take a moving frame E: U ⊂ M → Sp(2) along Φ. Note that E is also a moving frame along $\pi\circ\Phi$. It is not hard to see that using the U(1)×Sp(1)–action we can rechoose E so that $E^*\Sigma_0^1 = 0$ (Φ^{-1}Sp(2) → M is a U(1)×Sp(1)–principal bundle). This means that E is also a symplectic frame along $\pi\circ\Phi$. Now the horizontality of Φ gives $E^*\Sigma_0^0 = 0$. So the function b written relative to E is identically zero and the above lemma finishes the proof. □

Corollary. Any compact Riemann surface can be conformally and minimally immersed into $\mathbb{H}P^1 \equiv S^4$.

Proof. In §3 we have shown that any compact Riemann surface can be horizontally and holomorphically immersed (in fact embedded) into $\mathbb{C}P^3$. □

Given a conformal immersion f: M → $\mathbb{H}P^1$ and a symplectic frame along f put

$$Q = \bar{\sigma}_0^0 \cdot \sigma_1^1 \cdot \varphi \cdot \varphi.$$

We then have

Lemma. i) Q is a globally defined form on M, and

ii) Q is a type (4,0) holomorphic form on M if f is minimal.

We omit the proof of the above lemma as it is quite similar to the proofs of Lemma A–B of §2.

Remark. Suppose we are given a conformal minimal immersion f: M → $\mathbb{H}P^1$ with

$Q = Q_f \equiv 0$. Now $Q = -a\bar{c}(\varphi^1 + i\varphi^2)^4$ where a, c are local functions introduced in (14). Since the functions a and c are of analytic type (cf. Lemma C of §2) it follows that either $a \equiv 0$ (iff $\sigma_1^1 \equiv 0$) or $c \equiv 0$ (iff $\sigma_0^0 \equiv 0$). Also, given an immersion $f: M \to \mathbb{H}P^1$ observe that there exists a unique global lifting $\Phi_f: M \to \mathbb{C}P^3$ so that $\pi \circ \Phi_f = f$. We can now formulate the relationship between horizontal holomorphic curves in $\mathbb{C}P^3$ and conformal minimal surfaces in $\mathbb{H}P^1$ precisely as follows (the reader should have little difficulty in providing a proof by now): *A conformal minimal immersion $f: M \to \mathbb{H}P^1$ has $Q_f \equiv 0$ and $\sigma_0^0 \equiv 0$ if and only if $\Phi_f: M \to \mathbb{C}P^3$ is a horizontal holomorphic immersion.*

§5. Examples

Consider the map

(1) $$F: \mathbb{R}^3 = \{(x,y,z)\} \to \mathbb{R}^5 = \{(u^1,\cdots,u^5)\}$$

given by

$$u^1 = yz/\sqrt{3},\ u^2 = zx/\sqrt{3},\ u^3 = xy/\sqrt{3},\ u^4 = (x^2 - y^2)/2\sqrt{3},\ u^5 = (x^2 + y^2 - 2z^2).$$

$\Delta = -\frac{\partial}{\partial x^2} - \frac{\partial}{\partial y^2} - \frac{\partial}{\partial z^2}$ is the Laplace–Beltrami operator of $\mathbb{R}^3$ with the usual metric. We easily have $\Delta u^\alpha = 0$, $1 \leq \alpha \leq 5$. The rank of the Jacobian of F is three away from the origin. Hence F is a minimal immersion $\mathbb{R}^3\backslash\{0\} \to \mathbb{R}^5$. Let $u_0 = F(x_0,y_0,z_0)$ and let λ be an arbitrary nonnegative number. Then $F(\sqrt{\lambda}x_0, \sqrt{\lambda}y_0, \sqrt{\lambda}z_0) = \lambda u_0$ and it follows that $F(\mathbb{R}^3)$ is a cone.

Exercise. Let $S^n(r)$ be the n–sphere of radius r centered at the origin in $\mathbb{R}^{n+1}$. Also let $N \subset S^n(r)$ be an immersed submanifold of $S^n(r)$. Then N is minimal in $S^n(r)$ if and only if $\mathrm{Cone}(N) = \{\lambda p: p \in N, \lambda \geq 0\}$ is an immersed minimal submanifold of $\mathbb{R}^{n+1}$ away from the origin.

Now $F(\mathbb{R}^3) \cap S^4(1) = F(S^2(\sqrt{3}))$ and by the exercise above we obtain a minimal immersion

(2) $$f = F|_{S^2(\sqrt{3})}: S^2(\sqrt{3}) \to S^4 = S^4(1).$$

f is not an embedding. In fact, $f(x, y, z) = f(-x, -y, -z)$ and f projects down to give an embedding $\mathbb{R}P^2 \to S^4$. A little computation shows that f is an isometric immersion. It follows that

$$A(f) = 4\pi r^2 = 12\pi, \quad K \equiv \tfrac{1}{3}, \quad \tau^2 = \tfrac{1}{2}(1 - K) = \tfrac{1}{3}.$$

f is called the Veronese surface.

One can generalize the above example (cf. [Cher4]) to produce isometric immersions

$$S^2(r) \to S^{2m}(1), \text{ where } r^2 = m(m+1)/2.$$

For example, we get

(3) $$f: S^2(\sqrt{6}) \to S^6$$

from $F: \mathbb{R}^3 \to \mathbb{R}^7$ given by

$u^1 = \sqrt{6}\, z(-3x^2 - 3y^2 + 2z^2)/72$, $u^2 = x(-x^2 - y^2 + 4z^2)/24$, $u^3 = \sqrt{10}\, z(x^2 - y^2)/24$, $u^4 = \sqrt{15}\, x(x^2 - 3y^2)/72$, $u^5 = y(-x^2 - y^2 + 4z^2)$, $u^6 = \sqrt{10}\, xyz/12$, $u^7 = \sqrt{15}\, y(3x^2 - y^2)/72$. For this map we have $A(f) = 24\pi$, $K \equiv \frac{1}{6}$, $\tau_1^2 = \frac{5}{12}$, $\tau_2^2 = \frac{1}{4}$.

Exercise. Determine all conformal minimal immersions $\mathbb{C}P^1 \to S^{2m}$ with τ_i, $1 \le i \le m-1$, all constant.

Let $M = \mathbb{C}P^1 = \mathbb{C} \cup \{\infty\}$ and z, the affine coordinate. Put

$$g(z) = az, \quad f(z) = bz^{n+1}, \quad a,b \in \mathbb{C}\backslash\{0\},\ n \in \mathbb{Z}^+.$$

g and f are meromorphic functions on M with a pole at ∞. For each $n > 1$ we obtain a nonlinear horizontal holomorphic embedding $\Phi(f,g)$: $\mathbb{C}P^1 \to \mathbb{C}P^3$,

(4) $$\Phi(f,g) = {}^t[1,\ az,\ (1-n)bz^{n+1}/2,\ b(1+n)z^n/2a].$$

If $n = 2$ then $\Phi(f,g)(\mathbb{C}P^1) \subset \mathbb{C}P^3$ is a *rational normal curve.* By a theorem from §4 the maps $\Phi(f,g)$ in turn give confomal minimal immersions into S^4 upon projection $\mathbb{C}P^3 \to \mathbb{H}P^1 = S^4$.

Let $S^3 = \{(z, w) \in \mathbb{C}^2\colon |z|^2 + |w|^2 = 1\}$ and $T = \{(z, w)\colon |z|^2 = |w|^2 = \frac{1}{2}\}$. $T \subset S^3$ is called the *Clifford torus.* It is the unique non-equatorial minimal surface of constant curvature ($\equiv 0$) in S^3. Lawson generalized this example and constructed a compact embedded minimal surface of an arbitrary genus in S^3. In the following we will give a sketch of this construction. For details we refer the reader to [L1]. Let *st*: $S^3 \to \mathbb{R}^3 \cup \{\infty\}$ denote the stereographic projection coordinatizing S^3 away from the north pole. Under *st* the south pole corresponds to the origin $\in \mathbb{R}^3$ and the equatorial hypersphere of S^3 corresponds to the unit two-sphere in $\mathbb{R}^3$. We also note that the standard metric in S^3 becomes

$$ds^2 = 4(dx^2 + dy^2 + dz^2)/(1 + x^2 + y^2 + z^2)^2,$$

where $(x, y, z) \in \mathbb{R}^3$. From this we observe that straight lines through the origin and great circles of S^2 are geodesics. Fix $g \in \mathbb{Z}^+$. We construct a broken geodesic, denoted by γ_g, in $\mathbb{R}^3 = S^3\backslash\{\infty\}$ as follows: Start with two radii of $S^2 \subset \mathbb{R}^3$ lying in the xy–plane and close the curve by taking the shortest great circular arcs on S^2 from the end points of the radii to $(0,0,1) = S^2 \cap \{\text{the z–axis}\}$. From [L1] we have

Fact. There exists a conformal embedding (continuous at the boundary)

$$f_\gamma\colon \Delta = \text{the unit disc} \to S^3 = \mathbb{R}^3 \cup \{\infty\}$$

which represents a solution to Plateau's problem for the closed curve γ_g. Moreover, this surface $f_\gamma(\Delta)$ can be smoothly continued as a minimal surface by geodesic reflection across each of its boundary arcs. After a finite number of geodesic reflections $f_\gamma(\Delta)$ then yields a closed embedded surface of genus g in S^3.

Lawson also showed that a compact embedded minimal surface in S^3 divide S^3 into two diffeomorphic components, and conjectured that these two components are of equal volume. This conjecture was recently disproved by [K–P–S].

Chapter III. Holomorphic Curves and Minimal Surfaces in $\mathbb{C}P^n$

Much attention has been paid to the investigation of minimal surfaces in $\mathbb{C}P^n$ (and in more general spaces) in recent years. Inspired by the work of physicists Din and Zakrzewski [D–Z] Eells and Wood [E–W] first gave a rigorous mathematical treatment of the classification of minimal two–spheres in $\mathbb{C}P^n$. [R] later gave a classification of minimal two–spheres in the complex Grassmannian $\mathbb{C}G_{4,2}$. [U, C–W2] gave a description of minimal two–spheres in $\mathbb{C}G_{n,k}$ in general. [Gl, Y4] constructed a class of minimal surfaces in $\mathbb{H}P^n$.

Two features stand out in the works mentioned above (and in many other works not mentioned). One is the invocation of the Riemann–Roch theorem to say that certain symmetric holomorphic differentials vanish on S^2; the other is to classify minimal two–spheres in terms of holomorphic curves, or other holomorphic objects. Compact homomorphic curves in $\mathbb{C}P^n$ are well–understood objects in Algebraic Geometry. Indeed the space of all nondegenerate holomorphic two–spheres in $\mathbb{C}P^n$ of degree d is parametrized by $\mathbb{C}G_{d+1,n+1}$ and combining this with the classification theorem of [E–W] we see that the space of all nondegenerate minimal two–spheres in $\mathbb{C}P^n$ is parametrized by

$$\bigoplus_{d\geq n} (\mathbb{C}G_{d+1,n+1} \times \mathbb{Z}_{n+1}).$$

Consider a conformal minimal immersion f: $S^2 = \mathbb{C}P^1 \to S^{2m}$. Have

$$\pi\colon S^{2m} \to \mathbb{R}P^{2m}, \quad i\colon \mathbb{R}P^{2m} \hookrightarrow \mathbb{C}P^{2m},$$

where i is induced by the inclusion $\mathbb{R} \subset \mathbb{C}$. As the map i is totally geodesic the map $i\circ\pi\circ f\colon S^2 \to \mathbb{C}P^{2m}$ gives a totally real conformal minimal immersion. Therefore, as we will show in §4 of this chapter, to a minimal two–sphere in S^{2m} one can associate a holomorphic curve in $\mathbb{C}P^{2m}$, which is nothing but its directrix curve mentioned in §2 of Chapter II. Moreover, the minimal two–sphere can be reconstructed from its directix curve.

In §§1–2 we give a rather complete description of holomorphic curves in $\mathbb{C}P^n$. It is important to understand the holomorphic case as the bulk of known minimal surfaces in $\mathbb{C}P^n$ "comes" from holomorphic curves. Given a holomorphic curve h: $M \to \mathbb{C}P^n$ there arises a $U(1)^{n+1}$–principal bundle, $\mathcal{F} \to M$, and sections of this bundle, in turn, give rise to (generalized) minimal surfaces in $\mathbb{C}P^n$. This process is reversible, i.e., a minimal surface in $\mathbb{C}P^n$ is attached to some holomorphic curve via the above process only when $M \cong S^2$ or certain holomorphic symmetric differentials on M vanish. Often times such a minimal surface is called a superminimal surface or a pseudo–holomorphic curve.

The above mentioned process of producing minimal surfaces from a holomorphic curve can be generalized to manufacture a minimal surface from another minimal surface. Repeating the process one obtains an infinite sequence of (generalized) minimal surfaces in $\mathbb{C}P^n$. However, in general none of these surfaces thus obtained is (±)holomorphic.

No doubt an interesting remaining problem in the theory of minimal surfaces in complex projective space is to give a classification of higher genus minimal surfaces

§1. Hermitian Geometry and Singular Metrics on a Riemann Surface

Let N denote an n–dimensional complex manifold equipped with a hermitian metric

$$ds_N^2 = \Sigma\, \Theta^\alpha \otimes \bar{\Theta}^\alpha, \quad 1 \le \alpha \le n, \tag{1}$$

where each Θ^α is a local type (1,0) form on N. (Θ^α), collectively, are called a unitary coframe.

The type (1,0) metric connection ∇ on the holomorphic tangent bundle, $T^{(1,0)}N \to N$, is given by the skew–hermitian connection matrix (Θ^α_β) satisfying

$$d\Theta^\alpha = -\Theta^\alpha_\beta \wedge \Theta^\beta + \tau^\alpha,$$

where (τ^α) are type (2,0) forms, called the torsion forms.

Remark. The above ∇ is called a type (1,0) connection because its connection matrix written relative to a holomorphic coframe consists of type (1,0) forms. Of course, a unitary coframe is not holomorphic.

The Kähler form of N is defined to be $\frac{-1}{2}$ times the imaginary part of ds_N^2,

$$\frac{i}{2} \Sigma\, \Theta^\alpha \wedge \bar{\Theta}^\alpha.$$

(N, ds_N^2) is called a Kähler manifold if its Kähler form is closed.

Exercise. Show that N is Kähler if and only if the torsion forms vanish.

Hereafter, we assume that N is Kähler. We thus have

$$d\Theta^\alpha = -\Theta^\alpha_\beta \wedge \Theta^\beta. \tag{2}$$

Let (e_α) denote the unitary frame dual to (Θ^α). Then (2) becomes

$$\nabla e_\alpha = e_\beta \otimes \Theta^\beta_\alpha. \tag{3}$$

The curvature forms, (χ^α_β), are defined by

$$\chi^\alpha_\beta = d\Theta^\alpha_\beta + \Theta^\alpha_\gamma \wedge \Theta^\gamma_\beta. \tag{4a}$$

(N, ds_N^2) is said to be of constant holomorphic sectional curvature c if

$$\chi^\alpha_\beta = \frac{c}{4}\,(\Theta^\alpha \wedge \bar{\Theta}^\beta + \delta^\alpha_\beta\, \Sigma\, \Theta^\gamma \wedge \bar{\Theta}^\gamma). \tag{4b}$$

Example. Take $N = \mathbb{C}P^n$. $U(n+1)$ acts transitively on $\mathbb{C}P^n$ by

$$A\cdot[v] = [Av], \quad A \in U(n+1),\ v \in \mathbb{C}^{n+1}\backslash\{0\}.$$

The isotropy subgroup at $0 = {}^t[1,0,\cdots,0] \in \mathbb{C}P^n$ is $G_0 = U(1)\times U(n)$, where we include $G_0 \hookrightarrow U(n+1)$ by $(a,A) \mapsto \begin{pmatrix} a,0 \\ 0,A \end{pmatrix}$. Let $\mathfrak{m}$ denote the $Ad(G_0)$–invariant complementary subspace to $\mathfrak{g}_0$ (= the Lie algebra of G_0) given by

$$\mathfrak{m} = \left\{\begin{bmatrix} 0, & -{}^t\bar{X} \\ X, & 0 \end{bmatrix} : X \in \mathbb{C}^n\right\}.$$

We identify $\mathfrak{m}$ with $\mathbb{C}^n$ via $\begin{bmatrix} 0, & -{}^t\bar{X} \\ X, & 0 \end{bmatrix} \leftrightarrow X$. $\Omega = (\Omega^A_B)$, $0 \le A,B \le n$, denotes the $\mathfrak{u}(n+1)$–valued Maurer–Cartan form of $U(n+1)$. The $\mathfrak{m}$–component of Ω is given by

$$\Omega^\alpha_0 \otimes \epsilon_\alpha, \quad 1 \le \alpha \le n,$$

where $\epsilon_\alpha = {}^t(0,\cdots,0,1,0,\cdots,0) \in \mathbb{C}^n = \mathfrak{m}$. The forms (Ω^α_0) pull back to give type (1,0) forms on $\mathbb{C}P^n$

The Fubini–Study metric on $\mathbb{C}P^n$ normalized so that the holomorphic sectional curvature equals 4 is given by

(6) $$ds^2_N = \Sigma\, \omega^\alpha_0 \otimes \bar{\omega}^\alpha_0, \quad 1 \le \alpha \le n,$$

where $\omega^\alpha_0 = s^*\Omega^\alpha_0$ and s is a local section of $U(n+1) \to \mathbb{C}P^n$.

Index Convention. $0 \le A,B,C \le n$; $1 \le \alpha,\beta,\gamma \le n$.

$(\Theta^\alpha = \omega^\alpha_0)$ form a local unitary coframe on $\mathbb{C}P^n$. Using the Maurer–Cartan structure equations $d\Omega = -\Omega\wedge\Omega$ we compute that

$$d\Theta^\alpha = d\omega^\alpha_0 = s^* d\Omega^\alpha_0 = s^*(-\Omega^\alpha_\beta + \delta^\alpha_\beta\, \Omega^0_0) \wedge \omega^\beta_0.$$

It follows that

(7) $$\Theta^\alpha_\beta = s^*(\Omega^\alpha_\beta - \delta^\alpha_\beta\, \Omega^0_0).$$

We also compute that

$$d\Theta^\alpha_\beta = -s^*(\Omega^\alpha_\gamma \wedge \Omega^\gamma_\beta) + \omega^\alpha_0 \wedge \bar{\omega}^\beta_0 + \delta^\alpha_\beta\, \Sigma\, \omega^\gamma_0 \wedge \bar{\omega}^\gamma_0.$$

Therefore,

(8) $$\chi^\alpha_\beta = \omega^\alpha_0 \wedge \bar{\omega}^\beta_0 + \delta^\alpha_\beta \Sigma\, \omega^\gamma_0 \wedge \bar{\omega}^\gamma_0,$$

and $(\mathbb{C}P^N, ds^2_N)$ is of constant holomorphic sectional curvature 4.

Let M be a Riemann surface now. A hermitian metric on M can be written as

(9) $$ds^2 = h(z)\, dz \otimes d\bar{z},$$

where z is a local holomorphic coordinate. So $h(z) = \langle \partial/\partial z, \partial/\partial z\rangle_z > 0$. $\varphi = \sqrt{h}dz$ is a unitary coframe and (9) can be rewritten as

(10) $$ds^2 = \varphi \otimes \bar{\varphi}.$$

Write $\varphi = \varphi^1 + i\varphi^2$ for some real–valued 1–forms φ^1, φ^2. The underlying riemannian metric on M is

$$\varphi\cdot\bar{\varphi} \text{ (the symmetric product)} = (\varphi^1)^2 + (\varphi^2)^2.$$

(We often confuse $\varphi\cdot\bar{\varphi}$ with $\varphi\otimes\bar{\varphi}$ as one determines the other.) The Levi Civita connection form of $(M, \varphi\cdot\bar{\varphi})$, denoted by θ, is characterized by the equations

(11) $$d\varphi^1 = -\theta \wedge \varphi^2, \quad d\varphi^2 = \theta \wedge \varphi^1.$$

Put $\theta_c = -i\theta$. Then we can write

(12) $$d\varphi = -\theta_c \wedge \varphi,$$

i.e., θ_c is the complex connection form of $(M, \varphi\otimes\bar{\varphi})$.

The curvature form of $(M, \varphi\otimes\bar{\varphi})$ is given by

(13) $$\chi = d\theta_c + \theta_c \wedge \theta_c = d\theta_c.$$

The Gaussian curvature K of $(M, \varphi\cdot\bar{\varphi})$ satisfies the equation

(14) $$d\theta_c = \frac{K}{2}\, \varphi\wedge\bar{\varphi} = -iK\cdot\text{the Kähler form}.$$

Note that although φ is only locally defined ($\tilde{\varphi} = e^{it}\varphi$ is another unitary coframe) $\varphi\wedge\bar{\varphi}$ is globally defined and that the curvature form $d\theta_c$ is a globally defined type (1,1) form on M.

Let Δ denote the laplace–Beltrami operator of (M, $\varphi\cdot\bar{\varphi}$). We then have

(15) $$d\theta_C = \bar{\partial}\partial \log h = \frac{1}{4}\Delta \log h \; \varphi\wedge\bar{\varphi}.$$

Observe that a hermitian Riemann surface M is necessarily Kähler since $\dim_{\mathbb{C}} M = 1$.

We wish to consider hermitian metrics on M possibly with "analytic type" singularities. A motivation for this is that such singular metrics arise naturally as osculating metrics of a holomorphic curve in $\mathbb{C}P^n$.

Definition. Let U be a domain in M. A type (1,0) form ψ on U is said to be of analytic type if in a neighborhood of every point of U, ψ can be written as the product of an analytic type fuction (see Chapter II §2) and dz, where z is a local holomorphic coordinate. A positive *semi*definite hermitian inner product in the holomorphic tangent bundle $T^{(1,0)}M$ is called a *singular* hermitian metric if it can be given locally as

(16) $$ds^2 = \psi \otimes \bar{\psi},$$

where ψ is a, not identically zero, type (1,0) analytic type form. The *singular divisor* of (M, ds^2), denoted by D_S, is defined to be the zero divisor of ψ, i.e.,

(17) $$D_S = \sum_{p\in M} \mathrm{ord}_p(\psi)p.$$

Remark. D_S is a finite divisor if M is compact and $\deg D_S$ is the total number of zeros of ψ on M counted according to multiplicity. (Note that though ψ is only locally defined the zeros of ψ and their orders are well–defined on all of M.)

Let M be a Riemann surface equipped with a singular metric

$$\psi \otimes \bar{\psi} = h(z)\, dz \otimes d\bar{z}, \quad h(z) \geq 0.$$

We still have the following equations away from the support of D_S (= the zero set of ψ):

(18) $$d\psi = -\theta_c \wedge \psi; \quad \chi = d\theta_c = \frac{K}{2}\, \psi \wedge \bar{\psi}.$$

The following theorem generalizes the Gauss–Bonnet–Chern theorem and a proof can be found in [Y5] Chapter V §2.

Theorem. Let M be a compact Riemann surface of genus g equipped with a singular metric. Then

(19) $$\frac{i}{2\pi} \int_M \chi = 2 - 2g + \deg D_s.$$

§2. Holomorphic Curves in $\mathbb{C}P^n$

Let M be a Riemann surface and consider a holomorphic map

(1) $$f\colon M \to \mathbb{C}P^n.$$

To avoid redundant considerations we assume that f is *nondegenerate*, i.e., f(M) does not lie in a lower dimensional projective subspace.

The branch points of f (i.e., where $f_* = 0$) form an isolated set in M and as we saw in Chapter I f is a conformal minimal immersion away from the branch locus.

Assign M a Riemannian metric ds_M^2 from its conformal class and write

(2) $$ds_M^2 = (\varphi^1)^2 + (\varphi^2)^2 = \varphi \cdot \bar{\varphi},$$

where $\varphi = \varphi^1 + i\varphi^2$ is a local nonvanishing type (1,0) form.

Definition. A smooth map $e = (e_0, \cdots, e_n)\colon U \subset M \to U(n+1)$ is called a moving frame along f if

$$[e_0] = \pi \circ e = f,$$

where $\pi\colon U(n+1) \to \mathbb{C}P^n$ is the projection. To put it another way, a moving frame along f is a local section of the $U(1)\times U(n)$–principal bundle $f^{-1}U(n+1) \to M$.

Recall from §1 that the m–component of the Maurer–Cartan form of U(n+1) is given by

$$\Omega_m = (\Omega_0^\alpha) = \Omega_0^\alpha \otimes \epsilon_\alpha.$$

f^* pulls back type (1,0) forms on $\mathbb{C}P^n$ to type (1,0) froms on M and π^* gives an isomorphism between type (1,0) froms on $\mathbb{C}P^n$ and the $\mathbb{C}$–span of $\{\Omega_0^1, \cdots, \Omega_0^n\}$. It follows that if e is a moving frame along f then

(3) $$e^*\Omega_0^\alpha,\ 1 \le \alpha \le n, \text{ are all of type } (1,0).$$

Thus there exist smooth local complex–valued functions, Z_0^α, on M with

(4) $$e^*\Omega_0^\alpha = Z_0^\alpha\, \varphi.$$

Lemma. There is a moving frame e along f such that

(5) $$e^*\Omega_0^\lambda = 0 \text{ for } \lambda > 1.$$

Proof. The linear isotropy representation

$$i\colon U(1)\times U(n) \to GL(m)$$

is given by

$$i(e^{it}, A)\colon X \mapsto e^{it}AX, \quad X \in \mathbb{C}^n = m.$$

Let e, ẽ be two moving frames along f. On their commom domain they are related by ẽ = e·k, where k is a smooth map into U(1)×U(n). The lemma now follows from the transformation rule

(6) $$e^*\Omega_m = i(k)\, \tilde{e}^*\Omega_m. \quad \square$$

Let e: U ⊂ M → U(n+1) be as in the above lemma. Then the totality of moving frames along f making the pullbacks of Ω_0^λ, $\lambda > 1$, vanish is given by e·k, where k is a smooth local $U(1)^2\times U(n-1)$–valued function on M.

Notation. $e^*\Omega_B^A = \omega_B^A$.

Hereafter we choose e as in the lemma so that $\omega_0^1 = Z_0^1\varphi$, $\omega_0^\lambda = 0$, $\lambda > 1$.

Let $\tilde{e} = e \cdot k$ be another such moving frame along f, and write

$$\tilde{\omega} = \tilde{e}^* \Omega, \ \tilde{\omega}_0^1 = \tilde{Z}_0^1 \varphi.$$

Using (6) one verifies that

$$Z_0^1 \bar{Z}_0^1 = \tilde{Z}_0^1 \bar{\tilde{Z}}_0^1.$$

We put

(7) $$(r^1)^2 = Z_0^1 \bar{Z}_0^1, \ r^1 \geq 0.$$

$(r^1)^2$ is a globally defined smooth function on M.

Proposition. $(r^1)^2$ is a, not identically zero, analytic type function on M.

Proof. $r^1(p) = 0$ iff $p \in M$ is a branch point of f. f is assumed to be nondegenerate, in particular, nonconstant. Hence the zeros of r^1 are isolated. To show that $(r^1)^2$ is of analytic type it suffices to show that in a neighborhood Z_0^1 is. Exterior differentiate both sides of the equation $\omega_0^1 = Z_0^1 \varphi$ and obtain

$$[dZ_0^1 - Z_0^1(\theta_c + \omega_0^0 - \omega_1^1)] \wedge \varphi = 0,$$

where θ_c is the connection form of M with respect to φ. The result now follows from Chapter II §2 (14). □

Upon exterior differentiation the equations in (5) yield

$$\omega_1^\lambda \wedge \omega_0^1 = 0, \ \lambda > 1.$$

We thus conclude that the forms (ω_1^λ) are all of type (1,0). Computations with the reduced isotropy group ($= U(1)^2 \times U(n-1)$) action reveal that we can now choose a moving frame e along f so that in addition to (5) we also have

$$\omega_1^2 = Z_1^2 \varphi, \ \omega_1^{\lambda_1} = 0, \ \lambda_1 > 2.$$

Such moving frames are determined up to the structure group $U(1)^3 \times U(n-2)$. It is easily checked that $(r^2)^2 = Z_1^2 \bar{Z}_1^2$ ($r^2 \geq 0$) is a well–defined analytic type function on M. Again the nondegeneracy of f assures us that r^2 is not identically zero. Recursively proceeding we obtain

Theorem (Frenet Frames along a Holomorphic Curve). Let f: $M \to \mathbb{C}P^n$ be a nondegenerate holomorphic map, where M is a Riemann surface. Then there exists a $U(1)^{n+1}$–reduction, $\mathcal{F} \to M$, of the $U(1)\times U(n)$–principal bundle $f^{-1}U(n+1) \to M$ such that if e is any local section of $\mathcal{F} \to M$ then relative to e we have

$$(8) \qquad \omega_0^{\lambda_0} = \omega_1^{\lambda_1} = \cdots = \omega_{n-2}^{\lambda_{n-2}} = 0, \quad \lambda_i > i+1,$$

$$(9) \qquad \omega_0^1 = Z_0^1\varphi,\ \omega_1^2 = Z_1^2\varphi,\ \cdots,\ \omega_{n-1}^n = Z_{n-1}^n\varphi,$$

where $Z_{\alpha-1}^{\alpha}$, $1 \le \alpha \le n$, are all local complex–valued analytic type functions. Moreover, near a *regular* point of f (i.e., where $Z_0^1\cdots Z_{n-1}^n \neq 0$) we can further choose e so that (9) becomes

$$(10) \qquad \omega_0^1 = r^1\varphi,\ \cdots,\ \omega_{n-1}^n = r^n\varphi,$$

where $(r^\alpha)^2 = Z_{\alpha-1}^{\alpha}\bar{Z}_{\alpha-1}^{\alpha}$ $(r^\alpha > 0)$ are all globally defined not identically zero analytic type functions on M.

Remark. As a direct consequence of the above theorem we have a well–defined map

$$\Phi_f = (e_0,e_\alpha)/\sim: M \to U(n+1)/U(1)^{n+1}$$

given a holomorphic curve f: $M \to \mathbb{C}P^n$. We have a natural identification

$$U(n+1)/U(1)^{n+1} \cong F_{1,2,\cdots,n+1}(\mathbb{C}^{n+1}),$$

where $F_{1,2,\cdots,n+1}(\mathbb{C}^{n+1})$ is the space of flags in $\mathbb{C}^{n+1}$, called the complex flag manifold. With respect to the natural (i.e., U(n+1)–invariant) complex structure on $U(n+1)/U(1)^{n+1}$ the map Φ_f is holomorphic. For a detailed exposition of holomorphic curves in flag manifolds see [Y3].

We now present a slightly different perspective of the above construction. Begin with a nondegenerate holomorphic curve f: $M \to \mathbb{C}P^n$. In a neighborhood f can be holomorphically lifted to $\mathbb{C}^{n+1}\backslash\{0\}$. Let

$$v(z) = {}^t(v^0(z), \cdots, v^n(z))$$

be such a lifting, where z is a holomorphic coordinate on $U \subset M$. So the v^i's are holomorphic and $[v(z)] = f(z)$. Put

$$e_0 = v, \ \ e_1 = v' = \frac{dv}{dz}.$$

Suppose we have another lifting $\tilde{v}$ of f, $\tilde{v} = {}^t(\tilde{v}^0, \cdots, \tilde{v}^n)$. Then we must have, since $[\tilde{v}] = f$, $\tilde{v}^i = \lambda v^i$ for every i for some $\mathbb{C}^*$–valued function λ. Put

$$\tilde{e}_0 = \tilde{v}, \ \ \tilde{e}_1 = \tilde{v}'.$$

Now $\tilde{e}_1 = {}^t(\lambda' v^i + \lambda v^{i\prime}) = {}^t(\lambda' v^i) + \lambda e_1$. Therefore,

$$\tilde{e}_0 \wedge \tilde{e}_1 = \lambda^2 \, e_0 \wedge e_1$$

and the complex two–plane $\mathbb{C}$–span $\{e_0, e_1\} = [e_0 \wedge e_1]$ (assuming that $e_0 \wedge e_1 \neq 0$) is well–defined independent of the choice of a homogeneous lifting. Put

(11) $$f_1 = [e_0 \wedge e_1]: M \backslash \Sigma \to \mathbb{C}G_{n+1,2},$$

where Σ denotes the zero set of $e_0 \wedge e_1$ which is globally defined.

The Plücker embedding i: $\mathbb{C}G_{n+1,k+1} \hookrightarrow \mathbb{C}P^N$, $N = \binom{n+1}{k+1} - 1$, is defined as follows: Let $X = [E_0 \wedge \cdots \wedge E_k] \in \mathbb{C}G_{n+1,k+1}$, $E_i \in \mathbb{C}^{n+1}\backslash\{0\}$. Fix a basis $\{\epsilon_0, \cdots, \epsilon_n\}$ of $\mathbb{C}^{n+1}$ and write $E_0 \wedge \cdots \wedge E_k = \sum_\alpha P_{\alpha_0 \cdots \alpha_k} \epsilon_{\alpha_0} \wedge \cdots \wedge \epsilon_{\alpha_k}$. Then

$$i: X \mapsto [P_{\alpha_0 \cdots \alpha_k}].$$

Via the Plücker embedding $f_1: M\backslash\Sigma \to \mathbb{C}P^N$, $N = \binom{n+1}{2} - 1$.

Lemma. Let $H: U \subset M \to \mathbb{C}^{N+1}$ be a holomorphic map with isolated zeros at $\Sigma \subset U$ so that $[H] = h: U\backslash\Sigma \to \mathbb{C}P^N$ is a holomorphic map. Then there exists a unique holomorphic map $\hat{h}: U \to \mathbb{C}P^N$ with $\hat{h}|_{U\backslash\Sigma} = h$.

Proof. H: $z \mapsto {}^t(v^0(z), \cdots, v^N(z))$, v^i's holomorphic. Suppose $z_0 \in \Sigma$. So $v^i(z_0)$

$= 0$ for every i. Put $k = \min_i \operatorname{ord}_{z_0} v^i$. Consider the map $\hat{H}$: $z \mapsto {}^t(z^k v^0(z), \cdots, z^k v^N(z))$. Then $\hat{h} = [\hat{H}]$ does the job. □

Applying the lemma to $f_1|_{M\backslash\Sigma}$ we see that there exists a unique holomorphic extension of f_1 to all of M which we call once again f_1.

Definition. f_1: $M \to \mathbb{C}G_{n+1,2} = \mathbb{C}P^{n^*}$ ($\mathbb{C}G_{n+1,2}$, the Grassmann manifold of complex two–planes in $\mathbb{C}^{n+1}$, is identified with the space of projective lines in $\mathbb{C}P^n$) is called the dual curve of f or the first associated curve of f.

Maintaining the above notation let $e_0 = {}^t(v^i)$, $e_1 = {}^t(v^{i\prime})$, $e_2 = {}^t(v^{i\prime\prime})$. There are also the tilded quantities: $\tilde{e}_0 = {}^t(\tilde{v}^i) = {}^t(\lambda v^i)$, $\tilde{e}_1 = {}^t(\lambda' v^i + \lambda v^{i\prime})$, $\tilde{e}_2 = {}^t(\lambda'' v^i + 2\lambda' v^{i\prime} + \lambda v^{i\prime\prime})$. It follows that

$$[\tilde{e}_0 \wedge \tilde{e}_1 \wedge \tilde{e}_2] = [e_0 \wedge e_1 \wedge e_2]$$

whenever $e_0 \wedge e_1 \wedge e_2 \neq 0$ so that [ ] makes sense. The nondegeneracy assumption on f again guarantees that the zero set of $e_0 \wedge e_1 \wedge e_2$ is isolated. By the lemma the map $[e_0 \wedge e_1 \wedge e_2]$ is holomorphically extended to all of M.

Definition. $f_2 = [e_0 \wedge e_1 \wedge e_2]$: $M \to \mathbb{C}G_{n+1,3}$ is called the second associated curve of f.

Inductively proceeding we obtain

$$f_k = [e_0 \wedge \cdots \wedge e_k]: M \to \mathbb{C}G_{n+1,k+1} \subset \mathbb{C}P^N,$$

where $N = \binom{n+1}{k+1} - 1$, $0 \le k \le n-1$, $f_0 = f = [e_0]$.

We leave the proof of the following proposition as an exercise.

Proposition. A moving frame along f, $e = (e_0, \cdots, e_n)$: $U \subset M \to U(n+1)$, is a Frenet frame along f (that is, a local section of $\mathcal{F} \to M$) if and only if

$$[e_0 \wedge \cdots \wedge e_k] = f_k \text{ for every } k.$$

Theorem (Metric Structure Equations for a Holomorphic Curve). Let $f\colon M \to \mathbb{C}P^n$ be a nondegenerate holomorphic map from a Riemann surface (possibly noncompact) with a fixed metric $ds^2 = \varphi\cdot\bar\varphi$ in its conformal class. Then in $M\backslash\{\text{the zeros of } r^1, \cdots, r^n\}$ we have

$$\Delta \log r^i = K + 2(r^{i-1})^2 - 4(r^i)^2 + 2(r^{i+1})^2, \tag{12}$$

where $1 \le i \le n$, $r_0 = r_{n+1} = 0$, K is the Gaussian curvature of (M, ds^2), and Δ is the Laplace–Beltrami operator of (M, ds^2).

Proof. Upon exterior differentiation (10) yields

$$(d \log r^i - \theta_C - \omega_{i-1}^{i-1} + \omega_i^i) \wedge \varphi = 0, \quad 1 \le i \le n.$$

Now $d \log r^i$ is real and $-\theta_C - \omega_{i-1}^{i-1} + \omega_i^i$ is purely imaginary. Hence

$$*d \log r^i = i(\theta_C + \omega_{i-1}^{i-1} - \omega_i^i), \quad 1 \le i \le n.$$

From the Maurer–Cartan structure equations we get

$$d\omega_i^i = [(r^i)^2 - (r^{i+1})^2] \, \varphi \wedge \bar\varphi,$$

where we put $r^{i+1} = 0$. We also have $d\theta_C = \frac{K}{2}\, \varphi \wedge \bar\varphi$. Now

$$d*d \log r^i = \frac{i}{2}\, \Delta \log r^i \, \varphi \wedge \bar\varphi$$

and the result follows upon exterior differentiation. □

Corollary (Quantization Theorem). Let $f\colon M \to \mathbb{C}P^3$ be a holomorphic isometric immersion where M is a *compact* Riemann surface with a metric in its conformal class. Then the Gaussian curvature $K \ge \frac{4}{3}$ implies that $K \equiv \frac{4}{3}$.

Proof. $r^1 = 1$ since f is an isometric immersion. So the equations in (12) become

$$0 = K - 4 + 2(r^2)^2,$$

$$\Delta \log r^2 = K + 2 - 4(r^2)^2 + 2(r^3)^2,$$

$$\Delta \log r^3 = K + 2(r^2)^2 - 4(r^3)^2.$$

Combining we get

$$\Delta \log(r^2)^2 r^3 = 6(K - \tfrac{4}{3}).$$

Now $K \geq \frac{4}{3}$ says that $\Delta \log(r^2)^2 r^3 \geq 0$. So $\log(r^2)^2 r^3$ is a subharmonic function on M with singularities at the zeros of r^2 and r^3 where it goes to $-\infty$. Since M is compact $\log(r^2)^2 r^3$ then must attain a maximum in M. Hence it is constant by the maximum principle for subharmonic functions and $K \equiv \frac{4}{3}$. □

In the above one may replace $\mathbb{C}P^3$ by any $\mathbb{C}P^n$ and derive the corresponding quantization result. We leave this to the reader.

For the rest of this section we assume that M is a *compact* Riemann surface of genus g. Consider a nondegenerate holomorphic map $f: M \to \mathbb{C}P^n$. We know that each $f_k(M)$ is an *algebraic* curve (this is a special case of Chow's theorem) in $\mathbb{C}P^N$, $N = \binom{n+1}{k+1} - 1$. Put

$$d_k = \deg f_k(M) \subset \mathbb{C}P^N.$$

So d_k is the number of times $f_k(M)$ meets a generic hyperplane in $\mathbb{C}P^N$.

Let $p \in M$ and also let $\#(p)$ denote the ramification index of f at p. Using the inhomogeneous coordinates write

$$f(z) = {}^t[1, f^1(z), \cdots, f^n(z)].$$

Then $\#(p) = \min_\alpha \operatorname{ord}_p(f^\alpha)$. Put

$$\#_0 = \# = \sum_{p \in M} \#(p).$$

$\#_0$ is the total ramification index of f. Similarly define $\#_k(p)$ and $\#_k$ for $f_k: M \to \mathbb{C}P^N$.

Exercise. Show that $\#_k$ is the number of zeros of r^k counted with multiplicity.

Definition. The k–th osculating metric of f is a singular metric on M given by

$$ds_k^2 = \omega_k^{k+1} \otimes \bar{\omega}_k^{k+1}.$$

Its Kähler form is $\Lambda_k = \frac{i}{2}\, \omega_k^{k+1} \wedge \bar{\omega}_k^{k+1}$.

Let ds_N^2 denote the normalized Fubini–Study metric on $\mathbb{C}P^N$. A standard calculation shows that

$$f_k^* \, ds_N^2 = ds_k^2,$$

where f_k: $M \to \mathbb{C}P^N$ is the k–th associated curve of f. Therefore,

$$\Lambda_k = f_k^*(\text{the Kähler form of } (\mathbb{C}P^N, ds_N^2)).$$

Notation. $\varphi^k = \omega_k^{k+1}$.

Put

$$D_k = \text{the singular divisor of } ds_k^2$$

so that

$$\deg D_k = \#_k.$$

We now compute the connection form and the curvature form of (M, ds_k^2) relative to φ^k. Using the Maurer–Cartan structure equations of $U(n+1)$ we obtain

$$d\varphi^k = d\omega_k^{k+1} = (\omega_k^k - \omega_{k+1}^{k+1}) \wedge \varphi^k.$$

So $\theta_k = \omega_{k+1}^{k+1} - \omega_k^k$ is the complex connection form of (M, ds_k^2). Exterior differentiating again,

$$\begin{aligned} d\theta_k &= -\varphi^{k-1} \wedge \bar{\varphi}^{k-1} + 2\varphi^k \wedge \bar{\varphi}^k - \varphi^{k+1} \wedge \bar{\varphi}^{k+1} \\ &= 2i(\Lambda_{k-1} - 2\Lambda_k + \Lambda_{k+1}). \end{aligned}$$

We also have

$$d\theta_k = -i\, K_k \Lambda_k,$$

where K_k is the Gaussian curvature of $(M, \varphi^k \cdot \bar{\varphi}^k)$.

Theorem (Plücker Formulae). Let f: $M \to \mathbb{C}P^n$ be a nondegenerate holomorphic map from a compact Riemann surface M of genus g. Then

$$(13) \qquad 2g - 2 - \#_k = d_{k-1} - 2d_k + d_{k+1},$$

where $0 \le k \le n-1$, $d_{-1} = d_n = 0$, $d_0 = \deg f(M)$.

Proof. By Gauss–Bonnet–Chern

$$\frac{i}{2\pi}\int_M d\theta_k = 2 - 2g + \#_k.$$

Also

$$\int_M \Lambda_k = \text{the area of } (M, ds_k^2).$$

The Wirtinger theorem states that

$$d_k = \frac{1}{\pi}\cdot(\text{the area of } (M, ds_k^2)).$$

The result follows easily. □

Note that we could have integrated the metric structure equations given in (12) directly to obtain (13).

§3. Minimal Surfaces in a Kähler Manifold

Let N be a Kähler manifold with the hermitian metric

$$ds_N^2 = \sum_{\alpha=1}^{n} \Theta^\alpha \otimes \bar{\Theta}^\alpha,$$

where (Θ^α) are local type (1,0) forms on N. We consider a conformal immersion

$$(1) \qquad f\colon M \to N,$$

where M is a Riemann surface. The induced Riemannian metric on M is

$$ds^2 = \Sigma\, f^*\Theta^\alpha \cdot f^*\bar{\Theta}^\alpha$$

and by conformality of f we have

$$(2) \qquad ds^2 = \varphi\cdot\bar{\varphi}$$

for some nonvanishing type (1,0) local form φ on M.

Definition. A (local) unitary coframe (Θ^α) on N is said to be an *adapted* coframe along f if

(3) $$f^*\Theta^\lambda = 0 \text{ for } \lambda \geq 3.$$

For an adapted coframe (whose existence is easily eastablished) we have

$$f^*\Theta^\lambda = 0, \ \lambda \geq 3,$$
$$f^*\Theta^1 = X^1\varphi + Y^1\bar{\varphi}, \quad f^*\Theta^2 = X^2\varphi + Y^2\bar{\varphi}$$

for some local complex-valued functions X^i, Y^i on M.

Lemma. There exists an adapted coframe along f with

(4) $$f^*\Theta^1 = X\varphi, \quad f^*\Theta^2 = Y\bar{\varphi},$$

where $|X|^2 + |Y|^2 = 1$.

Proof. By (2) we have $\varphi\cdot\bar{\varphi} = \sum_{i=1}^{2} (X^i\varphi + Y^i\bar{\varphi})\cdot(\bar{X}^i\bar{\varphi} + \bar{Y}^i\varphi)$. It follows that

(†) $$|X^1|^2 + |X^2|^2 + |Y^1|^2 + |Y^2|^2 = 1,$$

(‡) $$X^1\bar{Y}^1 + X^2\bar{Y}^2 = 0.$$

(‡) says that the complex vectors ${}^t(X^i)$ and ${}^t(Y^i)$ are orthogonal with respect to the standard hermitian inner product in $\mathbb{C}^2$. Thus we may (and do) apply a unitary transformation to the pair (Θ^1, Θ^2) to make $X^2 = Y^1 = 0$. Put $X = X^1$, $Y = Y^2$. (†) then gives $|X|^2 + |Y|^2 = 1$. □

We now look at the underlying Riemannian structure of the Kähler manifold N. To do this we define real-valued 1-forms $(\theta^\alpha, \theta^{n+\alpha})$ by

(5) $$\Theta^\alpha = \theta^\alpha + i\theta^{n+\alpha}.$$

Then the underlying Riemannian metric of N is given by

(6) $$\Sigma\, \theta^\alpha\cdot\theta^\alpha + \theta^{n+\alpha}\cdot\theta^{n+\alpha} = \Sigma\, (\theta^\alpha)^2 + (\theta^{n+\alpha})^2$$

and $(\theta^\alpha, \theta^{n+\alpha})$ form an orthonormal coframe.

Define real-valued 1-forms θ^α_β, $\theta^{n+\alpha}_{n+\beta}$ by

(7) $$\Theta^\alpha_\beta = \theta^\alpha_\beta + i\theta^{n+\alpha}_{n+\beta},$$

where (Θ^α_β) are the complex connection forms of N relative to (Θ^α) so that

(8) $$d\Theta^\alpha = -\Theta^\alpha_\beta \wedge \Theta^\beta.$$

It follows that

(9) $$d\begin{bmatrix}\theta^\alpha \\ \theta^{n+\alpha}\end{bmatrix} = -\begin{bmatrix}\theta^\alpha_\beta & ,-\theta^{n+\alpha}_{n+\beta} \\ \theta^{n+\alpha}_{n+\beta} , & \theta^\alpha_\beta\end{bmatrix} \wedge \begin{bmatrix}\theta^\beta \\ \theta^{n+\beta}\end{bmatrix}, \text{ or simply}$$

$$d\begin{bmatrix}\theta^\alpha \\ \theta^{n+\alpha}\end{bmatrix} = -\theta_{LC} \wedge \begin{bmatrix}\theta^\beta \\ \theta^{n+\beta}\end{bmatrix}.$$

(Note that $\Theta^\alpha_\beta = -\bar{\Theta}^\beta_\alpha$ implies that $\theta^\alpha_\beta = -\theta^\beta_\alpha$, $\theta^{n+\alpha}_{n+\beta} = \theta^{n+\beta}_{n+\alpha}$.) θ_{LC} is the Levi–Civita connection matrix of N written relative to $(\theta^\alpha, \theta^{n+\alpha})$. From (9) we see that the underlying Riemannian structure of a Kähler manifold is quite special.

We wish to derive a condition for f in (1) to be minimal with respect to the underlying Riemannian metric of N. We begin with an adapted coframe, (Θ^α), along f. Rewrite (4) as

(4a) $$\bar{Y}\, f^*\Theta^1 - X\, f^*\bar{\Theta}^2 = 0.$$

Exterior differentiate both sides of (4a) using (8) and obtain

(10) $$[(Xd\bar{Y} - \bar{Y}dX) - X\bar{Y}(f^*\Theta^1_1 + f^*\Theta^2_2)] \wedge \varphi - f^*\Theta^1_2 \wedge \bar{\varphi} = 0.$$

Thus we may set

(11) $$(Xd\bar{Y} - \bar{Y}dX) - X\bar{Y}(f^*\Theta^1_1 + f^*\Theta^2_2) = a\varphi + b\bar{\varphi},$$

(12) $$-f^*\Theta^1_2 = b\varphi + c\bar{\varphi},$$

for some local complex-valued functions a, b, c on M.

Upon exterior differentiation (3) leads to

(13) $$X\, f^*\Theta_1^\lambda \wedge \varphi + Y\, f^*\Theta_2^\lambda \wedge \bar{\varphi} = 0,\ \lambda \geq 3.$$

Thus we are able to set

(14) $$X\, f^*\Theta_1^\lambda = a^\lambda\varphi + b^\lambda\bar{\varphi},$$

(15) $$Y\, f^*\Theta_2^\lambda = b^\lambda\varphi + c^\lambda\bar{\varphi},$$

for some local complex–valued functions a^λ, b^λ, c^λ.

Proposition. f in (1) is minimal if and only if

(16) $$b = b^\lambda = 0.$$

Proof. Let (Θ^α) be an adapted coframe along f chosen as in the above lemma. Then the corresponding Riemannian coframe looks like

$$f^*\theta^\lambda = f^*\theta^{n+\lambda} = 0,\ \lambda \geq 3,$$
$$f^*(\theta^1 + i\theta^{n+1}) = X\varphi,$$
$$f^*(\theta^2 + i\theta^{n+2}) = Y\bar{\varphi}.$$

We will come up with a Darboux coframe $(\tilde{\theta}^\alpha,\ \tilde{\theta}^{n+\alpha})$ along f by suitably modifying θ^1, θ^2, θ^{n+1}, θ^{n+2}. Replacing (Θ^1, Θ^2) by $(e^{ik_1}\Theta^1,\ e^{ik_2}\Theta^2)$ for suitably chosen k_1 and k_2 we can make X and Y real–valued. Since $|X|^2 + |Y|^2 = 1$ we may put, for some local function α,

$$f^*e^{ik_1}\Theta^1 = \cos\alpha\ \varphi,$$
$$f^*e^{ik_2}\Theta^2 = \sin\alpha\ \bar{\varphi}.$$

Notation. $e^{ik_i}\Theta^i = \hat{\Theta}^i = \hat{\theta}^i + i\hat{\theta}^{n+i}$.

Put

$$(\tilde{\theta}^1,\tilde{\theta}^2,\tilde{\theta}^{n+1},\tilde{\theta}^{n+2}) = (\hat{\theta}^1,\hat{\theta}^2,\hat{\theta}^{n+1},\hat{\theta}^{n+2})\begin{pmatrix}\cos\alpha & 0 & \sin\alpha & 0\\ \sin\alpha & 0 & -\cos\alpha & 0\\ 0 & \cos\alpha & 0 & \sin\alpha\\ 0 & -\sin\alpha & 0 & \cos\alpha\end{pmatrix},$$

$$\tilde{\theta}^i = \theta^i \text{ for } i \neq 1,\ 2,\ n+1,\ n+2.$$

Then

$$f^*\tilde{\theta}^\lambda = f^*\tilde{\theta}^{n+\lambda} = 0, \quad \lambda \geq 3,$$
$$f^*\tilde{\theta}^{n+1} = f^*\tilde{\theta}^{n+2} = 0.$$

That is to say, $(\tilde{\theta}^\alpha, \tilde{\theta}^{n+\alpha})$ is a Darboux coframe along f. Let II_λ, II_{n+1}, II_{n+2}, $II_{n+\lambda}$ denote the corresponding second fundamental forms. A standard calculation then shows that

$$II_{n+1} + iII_{n+2}$$
$$= (a + 2b + c)(\tilde{\theta}^1)^2 + 2i(a - c)\tilde{\theta}^1 \cdot \tilde{\theta}^2 + (-a + 2b - c)(\tilde{\theta}^2)^2$$

Likewise we find that

$$II_\lambda + iII_{n+\lambda}$$
$$= (a_\lambda + 2b_\lambda + c_\lambda)(\tilde{\theta}^1)^2 + 2i(a_\lambda - c_\lambda)\tilde{\theta}^1 \cdot \tilde{\theta}^2 + (-a_\lambda + 2b_\lambda - c_\lambda)(\tilde{\theta}^2)^2.$$

It follows that

$$\text{trace}\,(II_{n+1} + iII_{n+2}) = 4b,$$
$$\text{trace}\,(II_\lambda + iII_{n+\lambda}) = 4b_\lambda.$$

Now f is minimal iff $\text{trace}II_{n+1} = \text{trace}II_{n+2} = \text{trace}II_\lambda = \text{trace}II_{n+\lambda} = 0$, and the proposition follows. □

To illustrate our method let us consider a conformal minimal immersion

(17) $$f\colon M = \mathbb{C}P^1 \to \mathbb{C}P^3.$$

The normalized Fubini–Study metric on $\mathbb{C}P^3$ is given by the pullback of Ad(U(1)×U(3))–invariant product

$$\Sigma\, \Omega_0^\alpha \otimes \bar{\Omega}_0^\alpha, \quad 1 \leq \alpha \leq 3,$$

where $\Omega = (\Omega_B^A)$ is the Maurer–Cartan form of U(4). It follows that (use the Ad–invariance) the induced metric on M is given by

(18) $$ds^2 = \Sigma\, (e^*\Omega_0^\alpha)\cdot(e^*\bar{\Omega}_0^\alpha),$$

where e: U ⊂ M → U(4) is a moving frame along f.

Notation. $e^*\Omega = \omega$.

Lemma A. There exists a moving frame, e, along f such that

$$\omega_0^1 = X\varphi, \quad \omega_0^2 = Y\bar{\varphi}, \tag{19}$$

$$\omega_0^3 = 0, \tag{20}$$

where X and Y are local complex–valued functions with $|X|^2 + |Y|^2 = 1$. Moreover, if $\tilde{e}$ is any other moving frame with

$$\tilde{e}^*\Omega_0^1 = \tilde{X}\varphi, \quad \tilde{e}_0^*\Omega^2 = \tilde{Y}\bar{\varphi}, \quad \tilde{e}^*\Omega_0^3 = 0, \tag{21}$$

then $\tilde{e} = e\cdot k$, where $k\colon U \cap \tilde{U} \to U(1)^4$.

Proof. This follows from the following transformation rule:

$$\hat{e}^*\Omega = \mathrm{Ad}(g^{-1})e^*\Omega = g^{-1}(e^*\Omega)g, \tag{22}$$

where $\hat{e} = e\cdot g$ with g, a $U(1)\times U(3)$–valued local function. □

Lemma A gives a $U(1)^4$–reduction of the $U(1)\times U(3)$–principal bundle $f^{-1}U(4) \to M$.

Rewrite (19) as

$$\bar{Y}\omega_0^1 - X\bar{\omega}_0^2 = 0. \tag{19a}$$

Exterior differentiate both sides of (19a) using the Maurer–Cartan structure equations and obtain

$$[(Xd\bar{Y} - \bar{Y}dX) - X\bar{Y}(\omega_1^1 + \omega_2^2 - 2\omega_0^0)] \wedge \varphi - \omega_2^1 \wedge \bar{\varphi} = 0. \tag{23}$$

So for some local complex–valued functions a, b, c,

$$(Xd\bar{Y} - \bar{Y}dX) - X\bar{Y}(\omega_1^1 + \omega_2^2 - 2\omega_0^0) = a\varphi + b\bar{\varphi}, \tag{24a}$$

$$-\omega_2^1 = b\varphi + c\bar{\varphi}, \tag{24b}$$

and b = 0 by minimality.

Likewise (20), upon exterior differntiation, yields

$$X\omega_1^3 \wedge \varphi + Y\omega_2^3 \wedge \bar{\varphi} = 0 \tag{25}$$

and we set, for some complex valued a^3, b^3, c^3,

$$\text{(26)}\qquad X\omega_1^3 = a^3\varphi + b^3\bar{\varphi},\qquad Y\omega_2^3 = b^3\varphi + c^3\bar{\varphi}.$$

Again $b^3 = 0$ by minimality of f.

Lemma B. i) $\Lambda_1 = \omega_0^1\bar{\omega}_0^2\bar{\omega}_2^1$ is a globally defined symmetric type (3,0) form on M; ii) $\Lambda_2 = \omega_0^1\bar{\omega}_0^2\omega_1^3\bar{\omega}_2^3$ is a globally defined symmetric type (4,0) form on M.

Proof. Let e, $\tilde{e} = e\cdot k$ be chosen as in Lemma A and write $k = (e^{it_0}, e^{it_1}, e^{it_2}, e^{it_3})$ for some real-valued local functions $t_0, \cdots, t_3$. Using (22) we compute that

$$\text{(27)}\qquad \tilde{e}^*\Omega_b^a = e^{-it_a}e^{it_b}e^*\Omega_b^a,\quad 0 \le a,b \le 3.$$

From (27) we easily have

$$\tilde{\omega}_0^1\tilde{\bar{\omega}}_0^2\tilde{\bar{\omega}}_2^1 = \omega_0^1\bar{\omega}_0^2\bar{\omega}_2^1$$

showing that Λ_1 is well-defined. Similarly $\tilde{\Lambda}_2 = \Lambda_2$. □

Lemma C. i) $X\bar{Y}$: $U \subset M \to \mathbb{C}$ is a, not identically zero, analytic type function; ii) Λ_1 is holomorphic; iii) ω_2^1 vanishes identically; iv) X,Y: $U \subset M \to \mathbb{C}$ are both, not identically zero, analytic type functions.

Proof. Rewrite (24a) as

$$[d(X\bar{Y}) - X\bar{Y}(\omega_1^1 + \omega_2^2 - 2\omega_0^0 + 2dX/X)] \wedge \varphi = 0.$$

So $d(X\bar{Y}) \equiv X\bar{Y}\cdot$(complex-valued 1-form) (mod φ) and Chapter II §2 (14) proves i). To prove ii) we write, locally,

$$\Lambda_1 = -X\bar{Y}\bar{c}\lambda^3(dz)^3,$$

where $\varphi = \lambda dz$. So Λ_1 is holomorphic iff

$$\partial(X\bar{Y}\bar{c}\lambda^3)/\partial\bar{z} = 0 \text{ iff}$$

$$\text{(†)}\qquad d(X\bar{Y}\bar{c}\lambda^3) \equiv 0 \text{ (mod dz)}.$$

We have

(28) $$d\lambda \equiv -\lambda\theta_c \text{ (mod dz)}.$$

Also from (24a)

(29) $$d(X\bar{Y}) \equiv X\bar{Y}(\omega_1^1 + \omega_2^2 - 2\omega_0^0 + 2dX/X) \text{ (mod dz)}.$$

(24b) can be rewritten as $\omega_1^2 = \bar{c}\varphi$. Exterior differentiation of both sides of this equation leads to

(30) $$d\bar{c} - \bar{c}(\theta_c + \omega_1^1 - \omega_2^2) \equiv 0 \text{ (mod dz)}.$$

(†) now follows from (28–30) and we have proved ii). The Riemann –Roch theorem implies that there are no nonzero holomorphic differentials on $M = \mathbb{C}P^1$. Hence $\Lambda_1 = 0$. But

$$\Lambda_1 = (X\bar{Y})(\varphi)^2 \cdot \bar{\omega}_2^1$$

and $X\bar{Y}$ is a, not identically zero, analytic type function. iii) follows. We now exterior differentiate the equation $\omega_0^1 = X\varphi$ and use the fact that $\omega_2^1 = 0$ obtaining

$$[dX - X(\omega_0^0 - \omega_1^1 + \theta_c)] \wedge \varphi = 0$$

showing that X is of analytic type. Likewise considering the equation $\omega_0^2 = Y\bar{\varphi}$ one shows that Y is of anlaytic type. Finally observe that

$$X \equiv 0 \text{ iff } f|_U \text{ is antiholomorphic, and}$$
$$Y \equiv 0 \text{ iff } f|_U \text{ is holomorphic.}$$

So by assumption neither X nor Y can vanish everywhere. □

Lemma D. i) a^3, c^3: $U \subset M \to \mathbb{C}$ are analytic type functions; ii) Λ_2 is holomorphic; iii) $\Lambda_2 \equiv 0$, hence either $\omega_1^3 \equiv 0$ or $\omega_2^3 \equiv 0$.

Proof. Exterior differentiate both sides of $X\omega_1^3 = a^3\varphi$ and obtain

$$[da^3 - a^3(\theta_c + \omega_1^1 - \omega_3^3 + dX/X)] \wedge \varphi = 0.$$

This shows that a^3 is of analytic type. The case of c^3 is similar. Proofs of ii), iii) are omitted as they are totally similar to those of Lemma C ii), iii). □

Observation. $\omega_1^3 = 0$ *and* $\omega_2^3 = 0$ if and only if $f(M) \subset \mathbb{C}P^2$. So this case is excluded by the nondegeneracy assumption.

Summing up, we have

Theorem. Let $f\colon M = \mathbb{C}P^1 \to \mathbb{C}P^3$ be a nondegenerate not ±holomorphic conformal minimal immersion. Then there exists a $U(1)^4$–reduction, $\mathcal{F}_f \to M$, of the $U(1)\times U(3)$–principal bundle $f^{-1}U(4) \to M$ such that if e is any local section of $\mathcal{F}_f \to M$ then with respect to e we have

$$
\begin{gathered}
\omega_0^1 = X\varphi, \quad \omega_0^2 = Y\bar{\varphi}, \quad |X|^2 + |Y|^2 = 1, \\
\omega_0^3 = 0, \quad \omega_2^1 = 0, \\
\omega_1^3 = 0 \ \text{ or (not both) } \ \omega_2^3 = 0
\end{gathered}
\tag{31}
$$

Exercise. Determine all conformal minimal immersions $f\colon \mathbb{C}P^1 \to \mathbb{C}P^3$ with *constant* Gaussian curvature. (Hint: Exterior differentiate equations in (31) and obtain relations amongst the Gaussian curvature and extrinsic invariants.)

§4. Minimal Surfaces Associated to a Holomorphic Curve

Definition. Let M be a Riemann surface and also let (N, ds_N^2) be a *Riemannian* manifold. A smooth map $f\colon M \to N$ will be called a *generalized* conformal minimal immersion if the following three conditions are met: i) the set of branch points of f, denoted by Σ, is isolated; ii) $f|_{M\setminus\Sigma}$ is a conformal minimal immersion; iii) $f^*ds_N^2$ defines a singular metric on M.

Remark. A smooth map f from a Riemann surface M to a Riemannian manifold (N, ds_N^2) is said to be *weakly conformal* if locally

$$f^*ds_N^2 = \lambda\varphi\cdot\bar{\varphi},$$

where λ is a nonnegative function and φ is a type (1,0) nonvanishing form.

Given a smooth map $f: M \to (N, ds_N^2)$ the *energy* of f is defined to be

$$E(f) = \int_M \operatorname{trace}(f^* ds_N^2)\, dv,$$

where dv denotes the volume element of M with respect to a metric in its conformal class. (Ex. E(f) is well-defined.) A *harmonic* map is, by definition, a smooth map f which is a critical point of the energy functional *E*. It can be shown (cf. [G–O–R]) that a smooth map $f: M \to N$ is a generalized conformal minimal immersion if and only if it is a weakly conformal harmonic map.

Consider a nondegenerate holomorphic map

$$h: M \to \mathbb{C}P^n, \tag{1}$$

where M denotes a Riemann surface. From §2 we have the Frenet bundle

$$\mathcal{F}_h \to M \tag{2}$$

which is a $U(1)^{n+1}$–reduction of $h^{-1}U(n+1) \to M$.

Definition. Let $1 \le k \le n$. The k–th associated map (not to be confused with the k–th associated curve) of h is the smooth map given by

$$h_k = [e_k]: M \to \mathbb{C}P^n, \tag{3}$$

where $e = (e_0, \cdots, e_n)$ is a local section of $\mathcal{F}_h \to M$. (If e, $\tilde{e}$ are local sections of $\mathcal{F}_h \to M$ then $\tilde{e} = (e^{ia_0}e_0, \cdots, e^{ia_n}e_n)$, hence $[e] = [\tilde{e}]$.)

Theorem A. Let h be given as in (1). Then i) h_k, $1 \le k \le n-1$, are generalized conformal minimal immersions, neither holomorphic nor antiholomorphic; ii) h_n is antiholomorphic.

Proof. Let $\sigma \in \mathcal{S}_{n+1}$ = the permutation group on $\{0, \cdots, n\}$ and consider

$$E = (E_0, \cdots, E_n) = (e_{\sigma(0)}, \cdots, e_{\sigma(n)}) = e_\sigma,$$

where e is a Frenet frame along h, i.e., e is a local section of $\mathcal{F}_h \to M$. Let U_σ denote the elementary matrix with

$$E = e \cdot U_\sigma.$$

For example, if $E = (e_a, e_{a+1}, \cdots, e_n, e_0, \cdots, e_{a-1})$ then

$$U_\sigma = \begin{bmatrix} 0, & I_a \\ I_{n+1-a}, & 0 \end{bmatrix}.$$

We then have $E^*\Omega = \mathrm{Ad}(U_\sigma^{-1})e^*\Omega$, or

(4) $$E^*\Omega = U_\sigma^{-1}(e^*\Omega)U_\sigma.$$

We will now take $n = 3$, $k = 1$ and prove the theorem. The general case is completely similar, though notationally cumbersome. From §2 (8, 9) we have

(5) $$\begin{gathered} e^*\Omega_0^1 = Z^1\varphi, \quad e^*\Omega_1^2 = Z^2\varphi, \quad e^*\Omega_2^3 = Z^3\varphi, \\ e^*\Omega_0^2 = e^*\Omega_0^3 = e^*\Omega_1^3 = 0. \end{gathered}$$

Put $E = (e_1, e_2, e_0, e_3)$. Then $[E_0] = [e_1] = h_1$ and E is a moving frame along h_1. Applying (4) we obtain

(6) $$\begin{gathered} E^*\Omega_0^1 = Z^2\varphi, \quad E^*\Omega_0^2 = -\bar{Z}^1\bar{\varphi}, \quad E^*\Omega_1^3 = Z^3\varphi, \\ E^*\Omega_0^3 = E^*\Omega_1^2 = E^*\Omega_2^3 = 0. \end{gathered}$$

Consulting the equations in §3 (31) we conclude that h_1 is a conformal minimal immersion away from its branch points. (More directly, we have $E^*\Omega_1^2 = 0$ and $E^*\Omega_1^3 = Z^3\varphi$ showing that $b = b^3 = 0$.) The set of branch points of h_1 coincides with the zero set of $|Z^1|^2 + |Z^2|^2$. This set is isolated as we have already shown in §2 that the zero set of any Z^i is isolated. Finally we have

$$h_1^* ds_{FS}^2 = (|Z^1|^2 + |Z^2|^2)\varphi\cdot\bar{\varphi}$$

which is a singular metric on M since each $|Z^i|^2\varphi\cdot\bar{\varphi} = (r^i)^2\varphi\cdot\bar{\varphi}$ is a singular metric. □

Theorem B. Let $f\colon M = \mathbb{C}P^1 \to \mathbb{C}P^n$ be a nondegenerate conformal minimal immersion which is neither holomorphic nor antiholomorphic. Then there exists a holomorphic map

$$h_f\colon M = \mathbb{C}P^1 \to \mathbb{C}P^n$$

such that if $\mathcal{F}_h \to M$ is the Frenet bundle of h_f and $e^h = (e_0^h, \cdots, e_n^h)$ is a local section of this bundle, then

$$f = [e_k^h]$$

for some k, $1 \leq k \leq n-1$.

Proof. We will prove the theorem for $n = 3$. The general case is proved by writing down formulae generalizing §3 (31). We leave this as an exercise for the reader. Let e be a moving frame along f as in Lemma A of §3. Then from §3 (31) we have

$$\text{(7)} \qquad \begin{gathered} e^*\Omega_0^1 = X\varphi, \quad e^*\Omega_0^2 = Y\bar{\varphi}, \quad e^*\Omega_2^3 = c\bar{\varphi}, \\ e^*\Omega_0^3 = e^*\Omega_1^2 = e^*\Omega_1^3 = 0, \end{gathered}$$

where we have assumed without loss of generality that $e^*\Omega_1^3 = 0$. Recall that the totality of moving frames $\tilde{e}$ along f with

$$\tilde{e}^*\Omega_0^1 = \tilde{X}\varphi, \quad \tilde{e}^*\Omega_0^2 = \tilde{Y}\bar{\varphi}, \quad \tilde{e}^*\Omega_0^3 = 0$$

defines a $U(1)^4$–principal bundle over $M = \mathbb{C}P^1$ and thus the maps

$$[e_k]\colon \mathbb{C}P^1 \to \mathbb{C}P^3, \quad 0 \leq k \leq 3,$$

are well–defined. Consider $E = (E_0, E_1, E_2, E_3) = (e_3, e_2, e_0, e_1)$. Applying (4) we obtain

$$\text{(8)} \qquad \begin{gathered} E^*\Omega_0^1 = -\bar{c}\varphi, \quad E^*\Omega_1^2 = -\bar{Y}\varphi, \quad E^*\Omega_2^3 = X\varphi, \\ E^*\Omega_0^2 = E^*\Omega_0^3 = E^*\Omega_1^3 = 0. \end{gathered}$$

It follows that from §2 (8, 9) that $[e_3] = h_f$ is a holomorphic map and E is a Frenet frame along h_f. Put $e^h = E$ and get $f = [e_2^h]$. □

Remark. i) Theorem B generalizes without difficulty to include generalized conformal minimal immersions $\mathbb{C}P^1 \to \mathbb{C}P^n$.

ii) It is not necessary to assume that $M = \mathbb{C}P^1$: we need only to assume that certain holomorphic symmetric differentials (namely, Λ_1, Λ_2, and their higher order analogs) vanish on M.

We can summarize the content of Theorem A, B as follows. Given a nondegenerate holomorphic curve, h, in $\mathbb{C}P^n$ its Frenet bundle, $\mathcal{F}_h \to M$, gives rise to n–1 generalized minimal surfaces and an antiholomorphic curve. One simply takes a local section, $e = (e_0, e_k)$, of the Frenet bundle and the maps $[e_k]$. Conversely consider a generalized minimal surface, f, in $\mathbb{C}P^n$, where we assume that certain holomorphic symmetric diffferential forms vanish on it. Then there exists a "Frenet bundle" of f whose local sections are of the form

$$E = e_\sigma \quad \text{for some } \sigma \in \mathcal{S}_{n+1},$$

where e is a local section of the Frenet bundle of some holomorphic curve in $\mathbb{C}P^n$.

Example. i) Define $h: \mathbb{C}P^1 \to \mathbb{C}P^2$ by

$$[1, z] \mapsto [1, z, z^2], \quad \infty = [0, 1] \mapsto [0, 0, 1].$$

h is an holomorphic embedding and $h(\mathbb{C}P^1)$ is called the rational normal curve. From h we get a generalized conformal minimal immersion $h_1: \mathbb{C}P^1 \to \mathbb{C}P^2$ given by

$$[1, z] \mapsto [\bar{z} + 2z\bar{z}^2, z^2\bar{z}^2 - 1, -z^2\bar{z} - 2z], \quad \infty \mapsto [0, 1, 0].$$

ii) Let $w_1, w_2 \in \mathbb{C}\backslash\{0\}$ with $\mathrm{im}\,(w_1/w_2) > 0$. Also let $\Gamma = \mathbb{Z}w_1 \oplus \mathbb{Z}w_2 \subset \mathbb{C}$ denote the free abelian group generated by w_1, w_2. $\mathbb{C}/\Gamma$ is a complex torus: its complex structure is determined by the requirement that the projection $\mathbb{C} \to \mathbb{C}/\Gamma$ be holomorphic. (It is well–known that any compact Riemann surface of genus 1 arises in this fashion.) The Weierstrass $\wp$–function is the meromorphic function with double poles given by

$$\wp(z) = \frac{1}{z^2} + \sum_{w\in\Gamma\setminus\{0\}} [\frac{1}{(z-w)^2} - \frac{1}{w^2}], \quad z \in \mathbb{C}.$$

$\wp$ is doubly periodic with periods in Γ, hence projects down to give a meromorphic function, again denoted by $\wp$, on $\mathbb{C}/\Gamma$. There is the holomorphic embedding

$$h: \mathbb{C}/\Gamma \to \mathbb{C}P^2, \quad z \mapsto [1, \wp, \wp'].$$

Its (non ±holomorphic) associated map is given by

$$h_1: z \mapsto [\wp'\bar{\wp} + \wp''\bar{\wp}', \wp''\bar{\wp}'\wp - \wp' - \wp'^2\bar{\wp}', \wp'^2\bar{\wp} - \wp'' - \wp\bar{\wp}\wp''].$$

Remark. It is known that (see, for example, [Y5] Chapter III §5) the space of all nondegenerate holomorphic maps of degree d from $\mathbb{C}P^1$ into $\mathbb{C}P^n$ $(n \leq d)$ is naturally identified with $\mathbb{P}G(n, \mathbb{C}P^{d^*})$ = the space of projective n–planes in the space of hyperplanes of $\mathbb{C}P^d$. It follows that with fixed d, the space of all nondegenerate generalized conformal minimal immersions $\mathbb{C}P^1 \to \mathbb{C}P^n$ is parametrized by

$$\mathbb{P}G(n, \mathbb{C}P^{d^*}) \times \mathbb{Z}_{n+1}.$$

Consider a (generalized) conformal not ±holomorphic minimal immersion

$$f: M \to \mathbb{C}P^n,$$

where M is any Riemann surface. As in Lemma A of §3 there exists an adapted moving frame e along f so that

$$\omega_0^1 = X\varphi, \quad \omega_0^2 = Y\bar{\varphi}, \quad \omega_0^\lambda = 0, \; \lambda \geq 3,$$

where $\omega = e^*\Omega$. Such moving frames are determined up to the structure group $U(1)^3 \times U(n-2)$. So if $\tilde{e} = (\tilde{e}_0, \tilde{e}_1, \cdots, \tilde{e}_n)$ is another adapted moving frame along f then e and $\tilde{e}$ are related by $\tilde{e} = e\cdot k$, where k is a local $U(1)^3\times U(n-2)$–valued function. As a consequence we obtain well–defined maps

$$[e_1], [e_2]: M \to \mathbb{C}P^n.$$

Exercise. Show that the maps $[e_1]$, $[e_2]$: $M \to \mathbb{C}P^n$ are both generalized conformal minimal immersions.

The preceding discussion tells us that given a (generalized) minimal surface in $\mathbb{C}P^n$ we can manufacture, applying a process of differentiation repeatedly, an infinite family of (generalized) minimal surfaces in $\mathbb{C}P^n$. However, none of these manufactured surfaces are ±holomorphic in general.

Observe that $[e_3 \wedge \cdots \wedge e_n] = [\tilde{e}_3 \wedge \cdots \wedge \tilde{e}_n]$, where e and $\tilde{e}$ are any two adapted moving frames along a minimal surface f. Thus given a conformal minimal immersion f: $M \to \mathbb{C}P^n$ we also obtain a map

$$[e_3 \wedge \cdots \wedge e_n]\colon M \to \mathbb{C}G_{n+1,n-2}.$$

A natural question to ask is: Is the above map a generalized conformal minimal immersion with respect to the usual metric on $\mathbb{C}G_{n+1,n-2}$?

Chapter IV. Holomorphic Curves and Minimal Surfaces in the Quadric

Let G be a connected compact simple Lie group of rank n and also let $T \cong U(1)^n$ denote a maximal torus in G. G/T is naturally a Kähler manifold and it carries in its tangent bundle a rank n holomorphic distribution called the *horizontal distribution* (also called the superhorizontal distribution by some authors). Let H be a closed subgroup of G of maximal rank and further suppose that G/H is a type I inner symmetric space. An important theorem proved by Bryant [Br3] then states that: a horizontal curve in G/T projects down to G/H to give a generalized minimal surface. For G = SO(2n+1) of SO(2n) we thus obtain generalized minimal surfaces in S^{2n}, Q_{2n-1}, Q_{2n-2}.

For G = SO(4), G/T is the two–quadric in $\mathbb{C}P^3$ and we give a complete description of holomorphic curves and their associated minimal surfaces in §§1–2. For Q_m, $m > 2$, however, even the analysis of holomorphic curves is quite complicated (cf. [J–R–Y]). We shall be content with giving a thorough description of horizontal curves and associated minimal surfaces in SO(2n)/T and in SO(2n+1)/T, which we do in §§3–4.

§1. Immersed Holomorphic Curves in the Two–Quadric

Q_2 denotes the hyperquadric in $\mathbb{C}P^3$. (See Chapter I §3.) Q_2 is identified with $G_{n,2}$ = the Grassmann manifold of oriented two planes in $\mathbb{R}^4$:

$$[v + iw] \leftrightarrow [v \wedge w],$$

where $[v + iw]$ denotes the point in Q_2 given by the homogeneous vector $v + iw$ in $\mathbb{C}^4$ and $[v \wedge w]$ denotes the oriented two–plane in $\mathbb{R}^4$ spanned by the ordered pair v, w $\in \mathbb{R}^4$.

As a homogeneous space $Q_2 = SO(4)/SO(2)\times SO(2)$. Let $\Omega = (\Omega^\alpha_\beta)$ denote the Maurer–Cartan form of SO(4). Put

(1) $$\Omega^3 = \Omega^3_1 + i\Omega^3_2, \quad \Omega^4 = \Omega^4_1 + i\Omega^4_2.$$

Index Convention. $1 \leq \alpha,\beta,\gamma \leq 4$; $1 \leq i,j,k \leq 2$; $3 \leq a,b,c \leq 4$.

The Ω^a's pull back to give type (1,0) forms on M. The metric on Q_2 coming from the Fubini–Study metric on $\mathbb{C}P^3$ is given by

(2) $$\frac{1}{2} \Sigma\, s^*\Omega^a \cdot s^*\bar{\Omega}^a;$$

where s is a local section of $SO(4) \to Q_2$. We will call the above metric ds^2_{FS}.

Throughout this section we let M_1 be a Riemann surface and consider a holomorphic *immersion*

(3) $$f\colon M_1 \to Q_2.$$

Let $e = (e_\alpha)\colon U \subset M_1 \to SO(4)$ be a moving frame along f, i.e.,

$$[e_1 \wedge e_2] = [e_1 + ie_2] = f.$$

The holomorphy of f is reflected by the fact that $(e^*\Omega^a)$ are type (1,0) forms on M_1.

Notation. $e^*\Omega = \omega$.

The induced metric on M_1 can be written as

(4) $$f^* ds^2_{FS} = ds^2 = \varphi\cdot\bar{\varphi},$$

for some local type (1,0) nonvanishing form on M_1. Define local complex-valued functions X, Y by

(5) $$\omega^3 = X\varphi, \quad \omega^4 = Y\varphi.$$

(2) combined with (4) then implies

(6) $$X\bar{X} + Y\bar{Y} = 1.$$

Put $\tau_c = X^2 + Y^2$, $\tau = |X^2 + Y^2|$.

Lemma A. i) τ is a globally defined smooth function on M_1. ii) τ_c is of analytic type.

Proof. Let $\tilde{e}$ be another moving frame along f. Then on their common domain e and $\tilde{e}$ are related by

$$\tilde{e} = e \cdot k,$$

where $k = \left(\begin{pmatrix}\cos a, -\sin a \\ \sin a, \cos a\end{pmatrix}, \begin{pmatrix}\cos b, -\sin b \\ \sin b, \cos b\end{pmatrix}\right): U \cap \tilde{U} \to SO(2)\times SO(2)$. Define tilded quantities by

$$\tilde{e}^* \Omega^\alpha_\beta = \tilde{\omega}^\alpha_\beta, \quad \tilde{\omega}^a = \tilde{\omega}^a_1 + i\tilde{\omega}^a_2, \quad \tilde{\omega}^3 = \tilde{X}\varphi, \ \tilde{\omega}^4 = \tilde{Y}\varphi.$$

We then compute that

(7) $$\begin{bmatrix}\tilde{X} \\ \tilde{Y}\end{bmatrix} = \begin{bmatrix}\cos b, \sin b \\ -\sin b, \cos b\end{bmatrix} \begin{bmatrix}e^{-ia}X \\ e^{-ia}Y\end{bmatrix}.$$

It follows that

$$\tilde{X}^2 + \tilde{Y}^2 = e^{-2ia}(X^2 + Y^2)$$

showing that τ is well-defined globally. Note also that $0 \leq \tau \leq 1$. To prove ii) we assume that τ is not identically zero. Using the Maurer–Cartan structure equations of SO(4) we obtain

(8) $$\begin{aligned} d\omega^3 &= iX\omega^1_2 \wedge \varphi - Y\omega^3_4 \wedge \varphi, \\ d\omega^4 &= X\omega^3_4 \wedge \varphi + iY\omega^1_2 \wedge \varphi. \end{aligned}$$

We also have

(9) $$d\varphi = -\theta_c \wedge \varphi = i\theta \wedge \varphi,$$

where θ is the Levi–Civita connection form of $(M, \varphi\cdot\bar{\varphi})$. Exterior differentiate both sides of the equations in (5) using (8) and (9) and obtain

(10)
$$dX \equiv iX(\omega_2^1 - \theta) - Y\omega_4^3 \pmod{\varphi},$$
$$dY \equiv iY(\omega_2^1 - \theta) + X\omega_4^3 \pmod{\varphi}.$$

It follows at once that

(11) $$d\tau_c \equiv 2i\tau_c(\omega_2^1 - \theta) \pmod{\varphi}$$

showing that τ_c is of analytic type (see Chapter II §2 (14)). □

Lemma B. Suppose τ is identically zero. Then $f(M_1)$ is congruent to an open submanifold of totally geodesic $\mathbb{C}P^1 = U(2)/SO(2)\times SO(2) \subset Q_2$ with the Gaussian curvature 4. ($U(2)$ is included in $SO(4)$ via the identification $U(2) \leftrightarrow H = \{X \in SO(4): {}^tX\begin{pmatrix}J,0\\0,J\end{pmatrix}X = \begin{pmatrix}J,0\\0,J\end{pmatrix}$, where $J = \begin{pmatrix}0,-1\\1,\ 0\end{pmatrix}$.)

Proof. Put $Z = {}^t(X, Y)$, where X and Y are as in (5). Then $\tau \equiv 0$ implies that

$$\mathrm{Re}Z \perp \mathrm{Im}Z, \text{ and } |\mathrm{Re}Z| = |\mathrm{Im}Z| = 1/\sqrt{2}$$

at every point of M_1. Thus we can (and do) choose a moving frame $\tilde{e}$ about any point of M_1 with

$$\tilde{e}^*\Omega^3 = \varphi, \quad \tilde{e}^*\Omega^4 = i\varphi.$$

Consider the exterior differential system on $SO(4)$ given by

(12) $$\Sigma = \{\Omega_1^3 = \Omega_2^4,\ \Omega_2^3 = -\Omega_1^4\}.$$

Σ defines a completely integrable left–invariant distribution on $SO(4)$ whose analytic subgroup is $H = U(2)$. It is now an easy matter to check that there is a fixed $g \in SO(4)$ such that $g\cdot f(M_1)$ lies in $H/SO(2)^2$. □

For the rest of this section we assume that τ is not identically equal to zero. It is not hard to check that $\tau \equiv 0$ if and only if the induced Gaussian curvature is identically equal to four.

Notation. Z_τ = the zero set of τ.

In view of Lemma A, Z_τ is an isolated set in M_1.

Theorem A (Local Normal Form). Let f: $M_1 \to Q_2$ be a holomorphic immersion from a Riemann surface M_1 and define the invariant τ as in the above. Then in a neighborhood of every point of $M_1 \setminus Z_\tau$ there exists a moving frame, *e*, along f relative to which the following normal form holds:

$$\omega^3 = \cos\alpha\ \varphi, \quad \omega^4 = i\sin\alpha\ \varphi, \tag{13}$$

where α is a smooth local function with $-\pi/4 < \alpha < \pi/4$.

Proof. Let $p \in M_1 \setminus Z_\tau$. Then in a neighborhood of p, τ is never zero, hence τ_c, defined on a possibly smaller neighborhood, is never zero. Define real–valued *t by* $\tau_c = \tau e^{it}$. Possibly restricting to a yet smaller neighborhood we assume that *t* is a smooth function. Put $a = t/2$, $k = (e^{ia}, 1)$, and $\tilde{e} = e \cdot k$. Then using (7) we see that $\tilde{X}^2 + \tilde{Y}^2$ is real–valued. Putting $\tilde{Z} = {}^t(\tilde{X}, \tilde{Y})$ we then must have

$$\mathrm{Re}\tilde{Z} \perp \mathrm{Im}\tilde{Z}, \quad |\mathrm{Re}\tilde{Z}| > |\mathrm{Im}\tilde{Z}|.$$

Now we apply a rotation in the normal plane and change $\tilde{Z}$ to

$${}^t(a, ib), \text{ where } a > |b| \geq 0.$$

Since $a^2 + b^2 = 1$ by (6) we can find locally defined smooth α with

$$a = \cos\alpha,\ b = \sin\alpha, \quad -\pi/4 < \alpha < \pi/4. \quad \square$$

So, in $M_1 \setminus Z_\tau$ we have

$$\tau = \cos 2\alpha > 0. \tag{14}$$

Upon exterior differentiation we get

$$d\tau = -2\sin 2\alpha\ d\alpha. \tag{15}$$

On the other hand (11) gives

$$d\tau \equiv 2i\tau(\omega_2^1 - \theta) \pmod{\varphi}. \tag{16}$$

Combining (15) and (16) and using the fact that τ is real we obtain

$$2\tau(\theta - \omega_2^1) = {}^*d\tau, \tag{17}$$

where * is the Hodge operator of (M, $\varphi\cdot\bar{\varphi}$). Rewriting

(18) $$2(\theta - \omega_2^1) = {*}\mathrm{dlog}\ \tau.$$

Exterior differentiation of both sides of (18) leads to

(19) $$2(K - 2) = \Delta\log\ \tau,$$

where K is the Gaussian curvature, and Δ, the Laplace–Beltrami operator of (M, $\varphi\cdot\bar{\varphi}$). (19) holds in $M_1\backslash Z_\tau$.

Corollary A. Suppose τ is constant (and not zero). Then $\tau \equiv 1$. $f(M_1)$ is, then, congruent to an open submanifold of $S^2 = SO(3)/SO(2)\times\{1\} \subset Q_2$ with the Gaussian curvature 2, where SO(3) is included in SO(4) via $g \mapsto \binom{1,0}{0,g}$.

Proof. Using (13) the equations in (10) become

$$\mathrm{d}\cos\alpha \equiv -i\sin\alpha\ \omega_4^3 \pmod{\varphi},$$
$$i\mathrm{d}\sin\alpha \equiv \cos\alpha\ \omega_4^3 \pmod{\varphi}.$$

So $\omega_4^3 = 0$. Hence

$$0 = \mathrm{d}\omega_4^3 = \omega_1^3 \wedge \omega_1^4 + \omega_2^3 \wedge \omega_2^4 = -i\cos\alpha\cdot\sin\alpha\ \varphi \wedge \bar{\varphi} = 0.$$

Since $-\pi/4 < \alpha < \pi/4$ we must have $\sin\alpha = 0$, and $\alpha = 0$. Thus $\tau = 1$. The rest of the proof is left to the reader. (The exterior differential system to consider is $\{\Omega_1^4 = \Omega_2^4 = 0\}$ on SO(4).) □

Exercise. Suppose f: U (connected and open) $\subset M_1 \to Q_2$ is a *homogeneous* holomorphic embedding, i.e., there is a subgroup $H \subset SO(4)$ such that f(U) is congruent to an open submanifold of $H / H \cap SO(2)^2$. Then $\tau \equiv 0$ or $\tau \equiv 1$.

Corollary B. Let f be as in (3) and assume that M_1 is compact. Then $K \geq 2$ implies that $K \equiv 2$. (Keep in mind that we have excluded the case $K \equiv 4$ from our consideration.)

Proof. $K \geq 2$ combined with (19) implies that log τ is a subharmonic function with sigularities at Z_τ where it goes to $-\infty$. Since M_1 is assumed to be compact log τ attains a maximum in M_1, hence must be constant by the maximum principle for subharmonic functions. This finishes the proof. □

Theorem B [Y2]. Let f: $M_1 \to Q_2$ be a holomorphic immersion from a *compact* Riemann surface M_1 of genus g. Also assume that the Gaussian curvature K of M_1 with the induced metric is not identically equal to 4. Then

$$\#_\tau = 2(\tfrac{1}{\pi}\cdot A(f) - \chi), \tag{20}$$

where A(f) denotes the area of the immersion, $\chi = 2 - 2g$ = the Euler–Poincaré characteristic of M_1, and $\#_\tau$ denotes the number of zeros of τ counted with multiplicity.

Proof. We have

$$A(f) = \frac{i}{2}\int_{M_1} \varphi\wedge\bar{\varphi}.$$

By Gauss–Bonnet–Chern,

$$\chi = \frac{i}{4\pi}\int_{M_1} K\ \varphi\wedge\bar{\varphi}.$$

By the argument principle,

$$\frac{-i}{4\pi}\int_{M_1} \Delta\log\tau\ \ \varphi\wedge\bar{\varphi} = \#_\tau.$$

Integrate both sides of the equation in (19) over M_1. □

The following is an immediate corollary of the above theorem.

Corollary C. i) If M_1 is homeomorphic to S^2 then $A(f) \geq 2\pi$. (Remember that the case $K \equiv 4$ has been excluded.) Moreover, $A(f) = 2\pi$ if and only if $f(M_1)$ is congruent to $Q_1 = SO(3)/SO(2)$. ii) Suppose the genus of M_1 is positive. Then $\#_\tau$ is a *positive* even integer.

Given a holomorphic immersion f: M_1 (possibly noncompact) $\to Q_2$ we can define another smooth map $f^\perp$: $M_1 \to Q_2$ as follows: Let $e = (e_\alpha)$ be any moving frame along f and put

(21) $$f^{\perp} = [e_3 \wedge e_4]\colon M_1 \to Q_2.$$

$f^{\perp}(p)$ is just the oriented two–plane in $\mathbb{R}^4$ perpendicular to f(p).

Theorem C [Y2]. Let f: $M_1 \to Q_2$ be a holomorphic immersion from a Riemann surface M_1. We also assume that $f(M_1)$ does not lie in a totally geodesic $\mathbb{C}P^1 \subset Q_2$. Then $f^{\perp}\colon M_1 \to Q_2$ is a conformal not ±holomorphic minimal immersion isometric to f.

Proof. Let $e = (e_\alpha)$ be a (local) moving frame along f. Then

$$E = (E_\alpha) = (e_3, e_4, e_1, e_2)$$

is a moving frame along $f^{\perp}$.

Notation. $\overset{*}{E}\Omega = \hat{\omega}$, $\overset{*}{e}\Omega = \omega$.

We then have

$$\hat{\omega}^{\alpha}_{\beta} = \omega^{\gamma}_{\delta}, \text{ where } \gamma = \alpha + 2 \pmod 4 \text{ and } \delta = \beta + 2 \pmod 4).$$

Thus

(22) $$\begin{aligned} \hat{\omega}^3 &= \hat{\omega}^3_1 + i\hat{\omega}^3_2 = \omega^1_3 + i\omega^1_4, \\ \hat{\omega}^4 &= \hat{\omega}^4_1 + i\hat{\omega}^4_2 = \omega^2_3 + i\omega^2_4. \end{aligned}$$

The metric on M_1 induced by $f^{\perp}$ is

$$\tfrac{1}{2}(\hat{\omega}^3\cdot\bar{\hat{\omega}}^3 + \hat{\omega}^4\bar{\hat{\omega}}^4) = \tfrac{1}{2}(\omega^3\cdot\bar{\omega}^3 + \omega^4\bar{\omega}^4) = \varphi\cdot\bar{\varphi}$$

showing that $f^{\perp}$ is isometric to f. Suppose $f^{\perp}$ is holomorphic. Then the $\hat{\omega}^{a}$'s are type (1,0) and they can not both vanish. Without loss of generality we assume that $\hat{\omega}^3$ is never zero in a neighborhood. Consider $\hat{\omega}^3 + \omega^3 = i(\omega^1_4 + \omega^3_2)$. $\hat{\omega}^3$ and ω^3 are both of type (1,0) and this implies that $\omega^1_4 + \omega^3_2$ is of type (1,0). But $\omega^1_4 + \omega^3_2$ is purely real, hence we must have

(†) $$\omega^1_4 + \omega^3_2 = 0.$$

Likewise considering $\hat{\omega}^3 + i\omega^4 = \omega_3^1 + \omega_4^2$ we see that

$$(\ddagger) \qquad \omega_3^1 + \omega_4^2 = 0.$$

($\dagger$) and ($\ddagger$) together imply that $\omega^4 = -i\omega^3$ which in turn implies that $\tau = \tau_f \equiv 0$ contradicting our assumption that $f(M_1)$ does not lie in a totally geodesic $\mathbb{C}P^1 \subset Q_2$. Similarly $f^\perp$ antiholomorphic implies that $\tau \equiv 0$. From (22) we see that

$$\frac{1}{\sqrt{2}}(\omega_3^1, \omega_3^2, \omega_4^1, \omega_4^2)$$

is a local orthonormal coframe along the map $f^\perp$. Consider

$$\frac{1}{\sqrt{2}} \cdot A \cdot {}^t(\omega_3^1, \omega_3^2, \omega_4^1, \omega_4^2) = {}^t(\theta^1, \theta^2, \theta^3, \theta^4),$$

where A is given by

$$A = \begin{pmatrix} -c, & 0, & s, & 0 \\ 0, & -c, & 0, & s \\ 0, & s, & 0, & c \\ -s, & 0, & -c, & 0 \end{pmatrix}, \quad c = \cos\alpha, \ s = \sin\alpha.$$

Then $\theta^3 = \theta^4 = 0$, i.e., (θ^α) form a Darboux coframe along $f^\perp$. Have

$$d\theta^\alpha = -\theta^\alpha_\beta \wedge \theta^\beta \text{ and } \theta^a_i = h^a_{ij}\theta^j.$$

We will show that trace $h^a = 0$ for $a = 3, 4$. Routine computations show that

$$\theta^3_1 = \theta^4_2 = d\alpha, \quad \theta^3_2 = \tau\omega^4_3 = -\theta^4_1.$$

From this we obtain at once that

$$\text{trace } h^4 = 0 \text{ and } h^3_{11} = h^3_{22} = 0$$

finishing the proof. □

From the proof of Theorem C it is clear that the minimal surface $f^\perp$ is quite special. In fact the global symmetric quartic differential associated to $f^\perp$ given by

$$\Lambda = [(-h^3_{11} + ih^3_{12})^2 + (-h^4_{11} + ih^4_{12})^2]\cdot(\varphi)^4$$

vanishes. Call a minimal surface in Q_2 superminimal if its Λ vanishes. An interesting problem is to describe the totality of superminimal surfaces in Q_2 (and more generally in Q_n). We note that Q_2 is *not* of constant holomorphic sectional curvature, and as a consequence the symmetric quartic differential Λ is not necessarily holomorphic (cf. Chapter III §3 Lemma C). So a minimal two–sphere in Q_2 is not necessarily superminimal.

§2. Holomorphic Curves in Q_2

In this section we give a treatment of holomorphic curves (not necessarily immersed) in Q_2 from a slightly different perspective. The point of view presented in §§2–3 is expounded in [Y3] and the interested reader should consult it for more information.

Consider a nonconstant holomorphic map

(1) $$f\colon M \to Q_2 = SO(4)/SO(2)\times SO(2)$$

from a Riemann surface M.

Definition. We will say that *f* is *degenerate* if f(M), upon a congruence, lies in $U(2)/SO(2)^2$ or in $SO(3)/\{e\}\times SO(2)$, where $U(2), SO(3) \hookrightarrow SO(4)$ as in §1.

Assume for the rest of this section that f is nondegenerate. Put

$$E_{12} = \begin{pmatrix} 0 & 0 & 1 & 0 \\ 0 & 0 & 0 & 1 \\ -1 & 0 & 0 & 0 \\ 0 & -1 & 0 & 0 \end{pmatrix}, \quad F_{12} = \begin{pmatrix} 0 & 0 & 0 & -1 \\ 0 & 0 & 1 & 0 \\ 0 & -1 & 0 & 0 \\ 1 & 0 & 0 & 0 \end{pmatrix},$$

$$E'_{12} = \begin{pmatrix} 0 & 0 & 1 & 0 \\ 0 & 0 & 0 & -1 \\ -1 & 0 & 0 & 0 \\ 0 & 1 & 0 & 0 \end{pmatrix}, \quad F'_{12} = \begin{pmatrix} 0 & 0 & 0 & 1 \\ 0 & 0 & 1 & 0 \\ 0 & -1 & 0 & 0 \\ -1 & 0 & 0 & 0 \end{pmatrix}.$$

Also put

$$V_{12} = \mathbb{R}\text{-span}\ \{E_{12}, F_{12}\}, \quad V'_{12} = \mathbb{R}\text{-span}\ \{E'_{12}, F'_{12}\}.$$

Then

$$\mathfrak{o}(4) = \mathfrak{g}_0 \oplus V_{12} \oplus V'_{12},$$

where $\mathfrak{g}_0$ denotes the Lie algebra of $SO(2)^2$. $\mathfrak{m} = V_{12} \oplus V'_{12}$ is the orthogonal complement to $\mathfrak{g}_0$ with respect to the Killing form.

Put

$$v = xE_{12} + yF_{12} \leftrightarrow x + iy, \quad v' = xE'_{12} + yF'_{12} \leftrightarrow x + iy,$$

where we use the complex notation to write v and v′ relative to their respective bases. We then compute that

$$\mathrm{Ad}_k\colon v = x + iy \mapsto \exp(2\pi i(a-b))\cdot(x + iy),$$
$$v' = x + iy \mapsto \exp(2\pi i(a+b))\cdot(x + iy),$$

where $k = \left(\begin{pmatrix}\cos a, -\sin a\\ \sin a, \cos a\end{pmatrix}, \begin{pmatrix}\cos b, -\sin b\\ \sin b, \cos b\end{pmatrix}\right) \in SO(2)^2$. It follows that V_{12} and V'_{12} are the *root spaces* of SO(4) relative to the *maximal torus* $SO(2)^2$. V_{12} and V'_{12} generate holomorphic line bundles L_1 and L_2 over Q_2 and we obtain the Whitney sum decomposition

(2)
$$T^{(1,0)}Q_2 = L_1 \oplus L_2,$$

where $T^{(1,0)}Q_2 \to Q_2$ is the holomorphic tangent bundle over Q_2.

Let e: $U \subset M \to SO(4)$ be a moving frame along f and put

(3)
$$\varphi_1 = \tfrac{1}{2}\cdot[(\omega_3^1 + \omega_4^2) + i(\omega_3^2 - \omega_4^1)],$$
$$\varphi_2 = \tfrac{1}{2}\cdot[(\omega_3^1 - \omega_4^2) + i(\omega_3^2 + \omega_4^1)].$$

$\varphi_1 = \frac{1}{2}\cdot(-\omega^3 + i\omega^4)$ and $\varphi_2 = \frac{1}{2}\cdot(-\omega^3 - i\omega^4)$, hence the φ_i's are type (1,0) forms on M.

Lemma. Let f: $M \to Q_2$ be a nondegenerate holomorphic map. Then the forms φ_1 and φ_2 are not identically zero and of analytic type.

Proof. Exterior differentiate both sides of the equations in (3) using the Maurer–Cartan structure equations of SO(4). □

The induced *singular* metric on M is

$$ds^2 = \varphi_1 \cdot \bar{\varphi}_1 + \varphi_2 \cdot \bar{\varphi}_2 = \frac{1}{2}\,\Sigma(\omega_i^a)^2 \ (a = 3,4;\ i = 1,2). \tag{4}$$

Notation. $< \, , \, >_i = ds^2_{FS}|_{L_i}$.

We then have

$$f^* < \, , \, >_i = \varphi_i \cdot \bar{\varphi}_i. \tag{5}$$

We will call $ds_i^2 = \varphi_i \cdot \bar{\varphi}_i$, the *i–th osculating singular metric.* As in Chapter III §2 we put

$$\begin{aligned} D_i &= \text{the zero divisor of } \varphi_i \text{ (globally defined)}, \\ \#_i &= \deg D_i. \end{aligned} \tag{6}$$

Theorem. Let f: M → Q_2 be a nondegenerate holomorphic map from a compact Riemann surface M of genus g. Then

$$\#_i - 2g + 2 = 2d_i, \quad i = 1, 2, \tag{7}$$

where

$$d_i = \frac{i}{2\pi} \int_M \varphi_i \wedge \bar{\varphi}_i \ (= \text{the normalized area of } (M, ds_i^2)).$$

Proof. We have

$$d\varphi_i = -\omega_i \wedge \varphi_i, \tag{8}$$

where ω_i is the purely imaginary connection form of $(M, \varphi_i \otimes \bar{\varphi}_i)$. Exterior differentiation of both sides of the equations in (3) using (8) yields

$$\begin{aligned} [i(\omega_2^1 - \omega_4^3) + \omega_1] \wedge \varphi_1 &= 0, \\ [i(\omega_2^1 + \omega_4^3) + \omega_2] \wedge \varphi_2 &= 0. \end{aligned} \tag{9}$$

$i\omega_2^1$, $i\omega_4^3$, ω_i are all purely imaginary. Hence

(10) $$\omega_1 = -i(\omega_2^1 - \omega_4^3), \quad \omega_2 = -i(\omega_2^1 + \omega_4^3).$$

Using (10) and the Maurer–Cartan structure equations we obtain

(11) $$d\omega_2^1 = 2(\Lambda_1 + \Lambda_2), \quad d\omega_4^3 = 2(\Lambda_2 - \Lambda_1),$$

where $\Lambda_i = \frac{i}{2}\, \varphi_i \wedge \bar{\varphi}_i$ = the Kähler form of (M, ds_i^2). Exterior differentiation of both sides of the equations in (10), using (11), leads to

(12) $$K_i\Lambda_i = 4\Lambda_i, \quad i = 1, 2,$$

where K_i denotes the Gaussian curvature of (M, ds_i^2). Integrate the equation in (12) over M. □

If in the above theorem f is assumed to be immersive then comparing (7) to (20) of §1 we see that

(13) $$\#_1 + \#_2 = \#_\tau.$$

Theorem C of §1 easily generalizes to

Theorem. Let $f: M \to Q_2$ be a nondegenerate holomorphic map from a Riemann surface M. Then the normal map $f^\perp: M \to Q_2$ is a generalized conformal minimal immersion neither holomorphic nor antiholomorphic.

§3. Horizontal Holomorphic Curves in SO(m)–Flag Manifolds

In this section we give a complete description of so called horizontal curves in SO(m)–flag manifolds. Minimal surfaces associated to these curves include almost all known examples of minimal surfaces in S^n, $\mathbb{C}P^n$, Q_n. (Some notable exceptions are the compact minimal surfaces in S^3 given by Lawson [L1] and their generalizations recently given by Karcher, Pinkall and Sterling [K–P–S].)

We first consider $N = G/T$, where $G = SO(2n)$ and T is a maximal torus in G given by

$$T = \{\mathrm{diag}(D_1, \cdots, D_n): D_i = \begin{pmatrix} c_i, -s_i \\ s_i, \ c_i \end{pmatrix}, c_i = \cos 2\pi x_i, \text{ etc}\} = SO(2)^n.$$

G/T is also the manifold of full oriented even flags in $\mathbb{R}^{2n}$,

$$G/T = \tilde{F}_{2,4,\cdots,2n}(\mathbb{R}^{2n}).$$

Notation. We identify SO(2) with U(1) and write e^{ix} (or exp(ix)) instead of $\begin{pmatrix} \cos x, -\sin x \\ \sin x, \ \cos x \end{pmatrix}$.

There is the Lie algebra decomposition

(1) $$\mathfrak{o}(2n) = \mathfrak{t} \oplus \mathfrak{m},$$

where $\mathfrak{t}$ denotes the Lie algebra of T and $\mathfrak{m}$ is the vector space orhtogonal complement to $\mathfrak{t}$ relative to the Killing form.

Put E_{ij} = the 2n×2n matrix with 1 at (2i–1,2j–1) and (2i,2j)–entries, –1 at (2j–1,2i–1) and (2j,2i)–entries, and zeròs elsewhere; F_{ij} = the 2n×2n matrix with 1 at (2i,2j–1) and (2j,2i–1)–entries, –1 at (2i–1,2j) and (2j–1,2i)–entries, and zeros elsewhere; E'_{ij} = the 2n×2n matrix with 1 at (2i–1,2j–1) and (2j,2i)–entries, –1 at (2i,2j) and (2j–1,2i–1)–entries, and zeros elsewhere; F'_{ij} = the 2n×2n matrix with 1 at (2i–1,2j) and (2i,2j–1)–entries, –1 at (2j–1,2i) and (2j,2i–1)–entries, and zeros elsewhere. Then

(2) $$\mathfrak{m} = \Sigma\, V_{ij} \oplus V'_{ij} \ (1 \le i < j \le n),$$

where $V_{ij} = \mathbb{R}\text{–span}\,\{E_{ij}, F_{ij}\}$ and $V'_{ij} = \mathbb{R}\text{–span}\,\{E'_{ij}, F'_{ij}\}$.

For $t = \mathrm{diag}\,(D_1,\cdots,D_n) \in T$, $v = xE_{ij} + yF_{ij} \in V_{ij}$, and $v' = xE'_{ij} + yF'_{ij} \in V'_{ij}$ we compute that

(3) $$\begin{aligned} \mathrm{Ad}_t\colon\ & v \leftrightarrow x + iy \mapsto \exp(2\pi i(x_i - x_j))\cdot(x + iy), \\ & v' \leftrightarrow x + iy \mapsto \exp(2\pi i(x_i + x_j))\cdot(x + iy), \end{aligned}$$

where we use the complex notation to write v, v′ relative to their respective bases. (3) means that the root spaces of G relative to T are V_{ij}, V'_{ij} and the corresponding roots are

$$\Delta = \{\pm(x_i - x_j), \pm(x_i + x_j): 1 \leq i < j \leq n\}.$$

Notation. $\theta_{ij} = x_i - x_j$, $\theta'_{ij} = x_i + x_j$.

Write $\Delta = \Delta_+ \cup \Delta_-$, where $\Delta_+ = \{\theta_{ij}, \theta'_{ij}: 1 \leq i < j \leq n\}$. Δ_+ forms a system of *positive roots* and the resulting *simple roots* are

$$\Delta_S = \{\theta_{12}, \theta_{23}, \cdots, \theta_{n-1,n}, \theta'_{n-1,n}\}. \tag{4}$$

$\Omega = (\Omega^\alpha_\beta)$, $1 \leq \alpha,\beta \leq 2n$, denotes the $\mathfrak{o}(2n)$–valued Maurer–Cartan form of G. The Maurer–Cartan structure equations are

$$d\Omega^\alpha_\beta = -\Omega^\alpha_\gamma \wedge \Omega^\gamma_\beta.$$

For $1 \leq i < j \leq n$ put

$$\begin{aligned} \Theta^{ij} &= \tfrac{1}{2}\,[(\Omega^{2i-1}_{2j-1} + \Omega^{2i}_{2j}) + i(\Omega^{2i}_{2j-1} - \Omega^{2i-1}_{2j})], \\ \Theta'^{ij} &= \tfrac{1}{2}\,[(\Omega^{2i-1}_{2j-1} - \Omega^{2i}_{2j}) + i(\Omega^{2i}_{2j-1} + \Omega^{2i-1}_{2j})]. \end{aligned} \tag{5}$$

Then the canonical (G–invariant) complex structure on G/T is obtained by decreeing that the pullbacks of Θ^{ij}, Θ'^{ij} to G/T are all of type (1,0).

For an invariant metric on G/T we take the pullback of

$$\tfrac{1}{2}\,\Sigma\,(\Omega^{2i-1}_{2j-1})^2 + (\Omega^{2i}_{2j})^2 + (\Omega^{2i}_{2j-1})^2 + (\Omega^{2i-1}_{2j})^2, \quad 1 \leq i < j \leq n.$$

$ds^2_{G/T}$ will denote the above metric. $ds^2_{G/T} = ds^2_{FS}$ when $n = 2$.

Definition. The horizontal bundle, denoted by $\mathscr{H} \to G/T$, is the holomorphic subbundle of $T^{(1,0)}G/T \to G/T$ given by the following Pfaffian system on G/T:

$$\{\theta^{ij} = 0 \text{ unless } j = i+1;\ \theta'^{ij} = 0 \text{ unless } (i,j) = (n-1,n)\}, \tag{6}$$

where θ^{ij} denotes the pullback of Θ^{ij} to G/T by a local section of $G \to G/T$, etc.

We have the Whitney sum decomposition

$$\mathscr{H} = L_1 \oplus \cdots \oplus L_n, \tag{7}$$

where for $1 \leq i \leq n-1$, L_i is the holomorphic line bundle over G/T generated by the root space $V_{i,i+1}$ and L_n is the holomorphic line bundle generated by $V'_{n-1,n}$.

Definition. By a *horizontal curve in G/T* we will mean a holomorphic map from a Riemann surface M

$$h: M \to G/T \tag{8}$$

tangential to $\mathscr{H}$ (i.e., $h_* T^{(1,0)}M \subset \mathscr{H}$).

Take a horizontal curve $h: M \to G/T$ and also let $e: U \subset M \to G = SO(2n)$ be a moving frame along h. The holomorphy of h is reflected by the fact that the forms $e^*\Theta^{ij}$, $e^*\Theta'^{ij}$, $1 \leq i < j \leq n$, are all of type (1,0) on M.

Notation. $e^*\Theta^{ij} = \theta^{ij}$, $e^*\Omega^\alpha_\beta = \omega^\alpha_\beta$, etc.

The horizontality of h implies that

$$\{\theta^{ij} = 0 \text{ for } j \neq i+1,\ \theta'^{ij} = 0 \text{ for } (i,j) \neq (n-1,n)\}. \tag{9}$$

So the surviving type (1,0) forms on M are

$$\varphi_i = \theta^{i,i+1},\ 1 \leq i \leq n-1, \text{ and } \varphi_n = \theta'^{n-1,n}. \tag{10}$$

As in §2 we let

$$< , >_i = ds^2_{G/T} \text{ restricted to the line bundle } L_i.$$

Then we have

$$f^* < , >_i = \varphi_i \cdot \bar{\varphi}_i,\ 1 \leq i \leq n. \tag{11}$$

Definition. A horizontal curve $h: M \to G/T$ is said to be *nondegenerate* if h(M) does not lie in any H/H_0, where H is a connected closed subgroup of G of lower rank (as a connected compact Lie group) and $H_0 = H \cap T$.

Hereafter we exclude degenerate curves from our consideration to avoid redundant considerations.

Put

$$ds_i^2 = \varphi_i \cdot \bar{\varphi}_i. \tag{12}$$

Lemma. For a nondegenerate horizontal curve h in G/T the forms φ_i are of analytic type and not identically zero. Consequently each ds_i^2 is a singular metric on M.

We leave the proof of the above lemma as an exercise to the reader. (h is nondegenerate iff none of the φ_i's is identically zero. Also see p. 64 and p. 76 [Y3].)

Put

$$(13) \qquad \begin{aligned} D_i &= \text{the zero divisor of } \varphi_i \text{ on M}, \\ \#_i &= \deg D_i. \end{aligned}$$

Away from the zeros of φ_i we have

$$d\varphi_i = -\omega_i \wedge \varphi_i, \tag{14}$$

$$d\omega_i = \frac{1}{2} K_i \varphi_i \wedge \bar{\varphi}_i = (-iK_i) \cdot \Lambda_i, \tag{15}$$

where ω_i is the complex connection form, $d\omega_i$ is the curvature form, K_i is the Gaussian curvature, and Λ_i denotes the Kähker form of (M, $\varphi_i \otimes \bar{\varphi}_i$).

Exterior differentiating both sides of the equations in (9) we obtain

$$(16) \qquad \begin{aligned} &\omega^{2i-1}_{2j-1} = \omega^{2i}_{2j} = \omega^{2i}_{2j-1} = \omega^{2i-1}_{2j} = 0,\ j \neq i+1,\ 1 \le i \le n-2, \\ &\omega^{2i-1}_{2i+1} = \omega^{2i}_{2i+2},\ \omega^{2i}_{2i+1} = -\omega^{2i-1}_{2i+2},\ 1 \le i \le n-2. \end{aligned}$$

Using (16) we can rewrite the equations in (10) as follows.

$$(17)\qquad \begin{aligned} \varphi_i &= \omega_{2i+1}^{2i-1} + i\omega_{2i+1}^{2i}, \quad 1 \le i \le n-2, \\ \varphi_{n-1} &= \tfrac{1}{2}\,[(\omega_{2n-1}^{2n-3} + \omega_{2n}^{2n-2}) + i(\omega_{2n-1}^{2n-2} - \omega_{2n}^{2n-3})], \\ \varphi_n &= \tfrac{1}{2}\,[(\omega_{2n-1}^{2n-3} - \omega_{2n}^{2n-2}) + i(\omega_{2n-1}^{2n-2} + \omega_{2n}^{2n-3})]. \end{aligned}$$

Using (14), (17) and the Maurer–Cartan structure equations we differentiate both sides of the equations in (10) and obtain

$$(18)\qquad \begin{aligned} \omega_i &= -i(\omega_{2i}^{2i-1} - \omega_{2i+2}^{2i+1}), \quad 1 \le i \le n-1, \\ \omega_n &= -i(\omega_{2n-2}^{2n-3} + \omega_{2n}^{2n-1}). \end{aligned}$$

Using the Maurer–Cartan structure equations, (10), and the definition of the Λ_i's we also obtain

$$(19)\qquad \begin{aligned} d\omega_{2i}^{2i-1} &= -2\Lambda_{i-1} + 2\Lambda_i, \quad 1 \le i \le n-2, \ \Lambda_0 = 0, \\ d\omega_{2n-2}^{2n-3} &= -2\Lambda_{n-2} + 2(\Lambda_{n-1} + \Lambda_n), \\ d\omega_{2n}^{2n-1} &= -2\Lambda_{n-1} + 2\Lambda_n. \end{aligned}$$

Exterior differentiation of the equations in (18), using (19) together with (15), now gives

$$(20)\qquad \begin{aligned} K_i\Lambda_i &= -2\Lambda_{i-1} + 4\Lambda_i - 2\Lambda_{i+1}, \quad 1 \le i \le n-3, \ \Lambda_0 = 0, \\ K_{n-2}\Lambda_{n-2} &= -2\Lambda_{n-3} + 4\Lambda_{n-2} - 2\Lambda_{n-1} - 2\Lambda_n, \\ K_{n-1}\Lambda_{n-1} &= -2\Lambda_{n-2} + 4\Lambda_{n-1}, \\ K_n\Lambda_n &= -2\Lambda_{n-2} + 4\Lambda_n, \end{aligned}$$

where we assume $G = SO(2n)$ with $n \ge 5$.

For $n = 4$ we get

$$K_1\Lambda_1 = 4\Lambda_1 - 2\Lambda_2,$$
$$K_2\Lambda_2 = -2\Lambda_1 + 4\Lambda_2 - 2\Lambda_3 - 2\Lambda_4,$$
$$K_3\Lambda_3 = -2\Lambda_2 + 4\Lambda_3,$$
$$K_4\Lambda_4 = -2\Lambda_2 + 4\Lambda_4.$$

For $n = 3$ we get

$$K_1\Lambda_1 = 4\Lambda_1 - 2\Lambda_2 - 2\Lambda_3,$$
$$K_2\Lambda_2 = -2\Lambda_1 + 4\Lambda_2,$$
$$K_3\Lambda_3 = -2\Lambda_1 + 4\Lambda_3.$$

Finally for $n = 2$,

$$K_1\Lambda_1 = 4\Lambda_1,$$
$$K_2\Lambda_2 = 4\Lambda_2.$$

The last pair of equations duplicates (12) of §2.

Integrate both sides of the equations in (20) over M and obtain

$$\begin{aligned} &2g - 2 - \#_i = d_{i-1} - 2d_i + 2d_{i+1},\ 1 \le i \le n-3,\ \Lambda_0 = 0, \\ &2g - 2 - \#_{n-2} = d_{n-3} - 2d_{n-2} + d_{n-1} + d_n, \\ &2g - 2 - \#_{n-1} = d_{n-2} - 2d_{n-1}, \\ &2g - 2 - \#_n = d_{n-2} - 2d_n, \end{aligned} \tag{21}$$

where $d_i = \frac{1}{\pi}\int_M \Lambda_i$.

For the rest of this section we let $G = SO(2n+1)$. Include $SO(2n)$ in $SO(2n+1)$ via $g \mapsto \binom{1,0}{0,g}$. For a maximal torus in G we use $T = SO(2)^n$ in the above included in G which we again denote by T. (So T is really $\{e\}\times SO(2)^n$.) Generally speaking, we will identify an object in $SO(2n+1)$ (or in $\mathfrak{o}(2n+1)$) with its counterpart in $SO(2n)$ (or in $\mathfrak{o}(2n)$).

As the root structure of $SO(2n+1)$ is distinct from that of $SO(2n)$ it is necessary that we treat this case separately. Nevertheless the computations involved are similar and will mostly be suppressed.

As before we have the Lie algebra decomposition $\mathfrak{g} = \mathfrak{t} \oplus \mathfrak{m}$, $\mathfrak{m} = \mathfrak{t}^\perp$. Let E_{ij}, F_{ij}, E'_{ij}, F'_{ij} be as before except that they are now interpreted to be vectors in $\mathfrak{g} = \mathfrak{o}(2n+1)$. Also put E_{2i-1} = the $(2n+1)\times(2n+1)$ matrix with 1 at $(2i-1,2n+1)$–entry, -1 at $(2n+1,2i-1)$–entry, and zeros elsewhere; E_{2i} = the $(2n+1)\times(2n+1)$ matrix with 1 at $(2i,2n+1)$–entry, -1 at $(2n+1,2i)$–entry, and zeros elsewhere. Then

$$\mathfrak{m} = \Sigma\, V_{ij} \oplus V'_{ij} \oplus V_i = \mathfrak{o}(2n) \oplus \Sigma\, V_i \ (1 \le i < j \le n,\ 1 \le i \le n),$$

where $V_i = \mathbb{R}\text{–span}\,\{E_{2i-1}, E_{2i}\}$ and V_{ij}, V'_{ij} as before.

For $t = \text{diag}\,(1, D_1, \cdots, D_n) \in T$ and $v = xE_{2i-1} + yE_{2i} = \binom{x}{y}$ we compute that

$$\text{Ad}_t\colon v \leftrightarrow x + iy \to \exp(2\pi i x_i)\cdot(x + iy).$$

This together with the earlier computation shows that V_{ij}, V'_{ij}, V_i are the root spaces of G relative to T. The roots are

$$\Delta = \{\pm(x_i - x_j), \pm(x_i + x_j), \pm x_i\colon 1 \le i < j \le n\}.$$

Notation. $\theta_{ij} = x_i - x_j$, $\theta'_{ij} = x_i + x_j$, $\theta_i = x_i$.

For positive roots we take $\Delta_+ = \{\theta_{ij}, \theta'_{ij}, \theta_i\colon 1 \le i < j \le n\}$. The resulting simple roots are

$$\Delta_S = \{\theta_{12}, \theta_{23}, \cdots, \theta_{n-1,n}, \theta_n\}.$$

$\Omega = (\Omega^\alpha_\beta)$, $1 \le \alpha,\beta \le 2n+1$, denotes the $\mathfrak{o}(2n+1)$–valued Maurer–Cartan form of G. For $1 \leq i < j \leq n$ put

$$\Theta^{ij} = \frac{1}{2}[(\Omega^{2i-1}_{2j-1} + \Omega^{2i}_{2j}) + i(\Omega^{2i}_{2j-1} - \Omega^{2i-1}_{2j})],$$
$$\Theta'^{ij} = \frac{1}{2}[(\Omega^{2i-1}_{2j-1} - \Omega^{2i}_{2j}) + i(\Omega^{2i}_{2j-1} + \Omega^{2i-1}_{2j})],$$
$$\Theta^{i} = \Omega^{2i-1}_{2n+1} + i\Omega^{2i}_{2n+1}.$$

The pullbacks of Θ^{ij}, Θ'^{ij}, Θ^i to G/T are all of type (1,0) with respect to the canonical complex structure on G/T.

For an invariant metric on G/T we take the pullback of

$$\frac{1}{2}\Sigma\,[(\Omega^{2i-1}_{2j+1})^2 + (\Omega^{2i}_{2j})^2 + (\Omega^{2i}_{2j-1})^2 + (\Omega^{2i-1}_{2j})^2 + 2(\Omega^{2i-1}_{2n+1})^2 + 2(\Omega^{2i}_{2n+1})^2].$$

We consider a horizontal curve $h\colon M \to G/T$. Let $e\colon U \subset M \to G = SO(2n+1)$ be a moving frame along h and write $e^*\Omega = \omega$, $e^*\Theta = \theta$, etc. Then the horizontality of h gives

$$\{\theta^{ij} = 0 \text{ for } j \neq i+1;\ \theta'^{ij} = 0 \text{ for } 1 \leq i < j \leq n;\ \theta^i = 0 \text{ for } i \neq n\}.$$

The surviving type (1,0) forms on M are

$$\varphi_i = \theta^{i,i+1},\ 1 \leq i \leq n-1;\ \varphi_n = \theta^n = \omega^{2n-1}_{2n+1} + i\omega^{2n}_{2n+1}.$$

As before put

$$\#_i = \text{the degree of the zero divisor of } \varphi_i,$$
$$d_i = \frac{1}{\pi}\int_M \Lambda_i \quad \text{for } 1 \leq i \leq n-1,$$
$$d_n = \frac{1}{2\pi}\int_M \Lambda_n,$$

where Λ_i $(1 \leq i \leq n)$ denotes the Kähler form of $(M, \varphi_i \otimes \bar{\varphi}_i)$. Going through a derivation similar to the one already given earlier this section we obtain

$$2g - 2 - \#_i = d_{i-1} - 2d_i + d_{i+1},\ 1 \leq i \leq n,$$

where $d_0 = 0$, $d_{n+1} = d_n$.

§4. Associated Minimal Surfaces

Let $A = (A_1, \cdots, A_{2n}) \in G = SO(2n)$ and consider the projection

$$SO(2n) \to Q_{2n-2}, \quad A \mapsto [A_{2i-1} \wedge A_{2i}]$$

for $1 \le i \le n$. For each i this projection factors down to give a projection

$$(1) \qquad \pi_i\colon SO(2n)/SO(2)^n \to Q_{2n-2}.$$

Now let $A = (A_0, A_1, \cdots, A_{2n}) \in G = SO(2n+1)$ and consider projections

$$SO(2n+1) \to S^{2n}, \quad A \mapsto [A_0];$$

$$SO(2n+1) \to Q_{2n-1}, \quad A \mapsto [A_{2i-1} \wedge A_{2i}], \quad 1 \le i \le n.$$

These projections then factor down to give projections

$$(2) \qquad \begin{aligned} &\pi_0\colon SO(2n+1)/SO(2)^n \to S^{2n}, \\ &\pi_i\colon SO(2n+1)/SO(2)^n \to Q_{2n-1}. \end{aligned}$$

In the light of our discussion in this chapter we can rephrase Theorem 2 of Chapter II §2 as follows: *Let $f\colon \mathbb{C}P^1 \to S^{2n}$ be a conformal minimal immersion. Then there exists a horizontal curve $h_f\colon \mathbb{C}P^1 \to SO(2n+1)/SO(2)^n$ such that*

$$f = \pi_0 \circ h_f.$$

Similarly we may rephrase Theorem A–B of Chapter III §4 as follows: *Let $f\colon \mathbb{C}P^1 \to \mathbb{C}P^n$ be a conformal minimal immersion. Then there exists a horizontal curve $h_f\colon SO(2n+2)/SO(2)^{n+1}$ and some i such that $\pi_i \circ h_f(\mathbb{C}P^1)$ lies in a totally geodesic maximal linear subspace $\mathbb{C}P^n \subset Q_{2n}$ and*

$$f = \pi_i \circ h_f.$$

$$\begin{array}{ccc} U(n+1)/U(1)^{n+1} & \hookrightarrow & SO(2n+2)/SO(2)^{n+1} \\ \downarrow & & \downarrow \pi_i \\ \mathbb{C}P^n & \hookrightarrow & Q_{2n} \end{array}$$

In the other direction we have

Theorem. i) Let h: $M \to SO(2n)/SO(2)^n$ be a horizontal curve. Then

$$\pi_i \circ h: M \to Q_{2n-2}, \quad 1 \le i \le n,$$

are all generalized (possibly constant) conformal minimal immersions; ii) Given a horizontal curve h: $M \to SO(2n+1)/SO(2)^n$ we get generalized (possibly constant) conformal minimal immersions

$$\pi_0 \circ h: M \to S^{2n},$$
$$\pi_i \circ h: M \to Q_{2n-1}, \quad 1 \le i \le n.$$

We note that S^{2n} and Q_{2n-2} are both type I inner symmetric spaces and the above theorem follows essentially from a theorem of Bryant [Br3]. The reader may produce a direct proof using the method in the proof of Theorem C §1 of this chapter.

§5. Minimal Surfaces in the Quaternionic Projective Space

In this section we give a complete description of horizontal holomorphic curves in Sp(n)/T. The full list of type I inner symmetric spaces G/H with G = Sp(n) are Sp(n)/U(n), Sp(n)/Sp(p)×Sp(n–p). In particular, a horizontal holomorphic curve in Sp(n)/T, upon projection, gives a generalized minimal surface in the quaternionic projective space $\mathbb{H}P^{n-1} = Sp(n)/Sp(1)\times Sp(n-1)$.

Let G = Sp(n) and also let $T = U(1)^n$. (We think of the quaternions $\mathbb{H}$ as $\{z_1 + z_2 j: z_i \in \mathbb{C}\}$. This induces the inclusion $U(1)^n \subset Sp(n)$.) Recall that

$$Sp(n) = \{X \in GL(n;\mathbb{H}): {}^t\bar{X}X = I\},$$

where – denotes the quaternionic conjugation, i.e., it sends $(z_1 + z_2 j)$ to $(\bar{z}_1 - z_2 j)$. T is a maximal torus in G and G/T is the Sp(n)–flag manifold of real dimension

$2n^2$. There is the fibration $G/T \to Sp(n)/Sp(1)^n = F_{1,2,\cdots,n}(\mathbb{H}^n)$ with the standard fibre $(Sp(1)/U(1))^n = \mathbb{C}P^1\times\cdots\times\mathbb{C}P^1$. $F_{1,2,\cdots,n}(\mathbb{H}^n)$ is the full quaternionic flag manifold in $\mathbb{H}^n$.

We have the usual decomposition $\mathfrak{g} = \mathfrak{t} \oplus \mathfrak{m}$. $\mathfrak{t}$, the Lie algebra of T, consists of purely imaginary (real multiples of i) $n\times n$ diagonal matrices. $\mathfrak{m}$ is the orthogonal vector space complement (with respect to the Cartan–Killing form) to $\mathfrak{t}$. Let E_{ab} = the $n\times n$ matrix with +1 at (a,b)–entry, –1 at (b,a)–entry and zeros elsewhere. Here $1 \le a < b \le n$. Also let F_{ab} = the $n\times n$ matrix with +1 at (a,b) and (b,a)–entry and zeros elsewhere. Here $1 \le a \le b \le n$.

Put

$$V_{ab} = \mathbb{R}E_{ab} \oplus i\,\mathbb{R}F_{ab},\quad 1 \le a < b \le n,$$
$$V'_{ab} = j\mathbb{R}F_{ab} \oplus k\,\mathbb{R}F_{ab},\quad 1 \le a \le b \le n.$$

Then

$$\mathfrak{m} = \oplus\Sigma\, V_{ab}\ (1 \le a < b \le n) \oplus\Sigma\, V'_{ab}\ (1 \le a \le b \le n).$$

Let $S(n;\mathbb{C})j$ denote the set of all complex symmetric $n\times n$ matrices multiplied on the right by j (j is a unit quaternion). Then we can write

$$\mathfrak{m} = (\mathfrak{u}(n)\backslash\mathfrak{t}) \oplus S(n;\mathbb{C})j.$$

Let $t = (e^{it_a}) \in T = U(1)^n$. The adjoint representation of T on $\mathfrak{m}$ is given by

$$Ad_t\colon v_{ab} = xE_{ab} + iyF_{ab}\ (\in V_{ab}) \mapsto e^{i(t_a - t_b)}v_{ab},$$
$$v'_{ab} = xF_{ab}j + iyF_{ab}j\ (\in V'_{ab}) \mapsto e^{i(t_a + t_b)}v'_{ab}.$$

Now $\mathfrak{t} \cong \mathbb{R}^n$ via $\mathrm{diag}(ix_1,\cdots,ix_n) \mapsto (x_1,\cdots,x_n)$. Then the roots $(\subset \mathfrak{t}^*)$ are

$$\{\pm 2x_a,\ \pm(x_a + x_b),\ \pm(x_a - x_b)\colon 1 \le a < b \le n\}.$$

For positive roots we take $\{2x_a,\ x_a + x_b,\ x_a - x_b\colon 1 \le a < b \le n\}$. The resulting simple roots are given by

$$\Delta_s = \{x_1 - x_2,\ x_2 - x_3,\ \cdots,\ x_{n-1} - x_n,\ 2x_n\}.$$

$\Omega = (\Omega^a_b)$ denotes the $\mathfrak{sp}(n)$–valued Maurer–Cartan form of G. We note that $\mathfrak{sp}(n)$ consists of n×n $\mathbb{H}$–valued matrices with ${}^t\bar{X} = -X$, hence $\Omega^a_b = -\bar{\Omega}^b_a$.

Let Γ^a_b, Σ^a_b be the complex–valued 1–forms with

$$\Omega^a_b = \Gamma^a_b + \Sigma^a_b j.$$

Thus Γ is $\mathfrak{u}(n)$–valued and Σ is $S(n;\mathbb{C})$–valued.

The Maurer–Cartan structure equations $d\Omega = -\Omega \wedge \Omega$ become

$$d\Gamma = -\Gamma \wedge \Gamma + \Sigma \wedge \bar{\Sigma}, \quad d\Sigma = -\Gamma \wedge \Sigma - \Sigma \wedge \bar{\Gamma}.$$

The 1–forms Γ^a_b $(1 \leq a < b \leq n)$ and Σ^a_b $(1 \leq a \leq b \leq n)$ pull back to give type (1,0) forms on G/T.

For an invariant metric on G/T we take the invariant metric corresponding to the inner product on $\mathfrak{m} = T_T G/T$ given by $-1/(2n+2)$ times the Cartan–Killing form restricted to $\mathfrak{m}$. (This metric is used in Corollary given at the end of this section.)

We consider a holomorphic map

$$f\colon M \to G/T,$$

where M is a Riemann surface. A local C^∞ lifting $e = (e_1, \cdots, e_n)$ into Sp(n) of f will be called a symplectic frame along f. If $\tilde{e} = (\tilde{e}_a)$ is any other symplectic frame along f then on their common domain we must have $\tilde{e} = e \cdot t$, where t is a T–valued C^∞ local function on M. Thus there are well–defined maps

$$[e_a]\colon M \to S^{4n-1}/U(1),$$

where S^{4n-1} = the unit vectors in $\mathbb{H}^n$.

Choose a symplectic frame e and put

$$\omega = e^*\Omega, \ \gamma = e^*\Gamma, \ \sigma = e^*\Sigma.$$

The holomorphy of f implies that

(†) $\quad \gamma^a_b$ $(1 \leq a < b \leq n)$, σ^a_b $(1 \leq a \leq b \leq n)$ are all of type (1,0).

From now on assume that f is also horizontal. Consulting the simple roots Δ_S we see that we then must have

(‡) $$\gamma^a_b = 0, \quad 1 \le a < b \le n,\ b \ne a+1; \quad \sigma^a_b = 0, \quad (a,b) \ne (n,n).$$

Put

$$\varphi_a = \gamma^a_{a+1},\ 1 \le a \le n-1,\ \varphi_n = \sigma^n_n.$$

Also put

$$\Lambda_a = \frac{i}{2}\, \varphi_a \wedge \bar{\varphi}_a.$$

A horizontal holomorphic map f: M → G/T is said to be *degenerate* if f(M) is congruent to an open submanifold of some G'/T∩G', where G' is a lower rank subgroup. A proof of the following proposition can be found in [Y4] p. 271.

Proposition. Let f: M → Sp(n)/T be a horizontal holomorphic map. Then f is nondegenerate if and only if none of the Λ_a's are identically zero.

It is routinely checked that for a nondegenerate horizontal curve each

$$ds_a = \varphi_a \bar{\varphi}_a \quad (1 \le a \le n)$$

defines a singular metric on M. The Kähler form for ds_a is Λ_a.

Assume that f is a nondegenerate horizontal curve. Using the Maurer–Cartan structure equations and (‡) we obtain

$$d\varphi_a = -(\gamma^a_a - \gamma^{a+1}_{a+1}) \wedge \varphi_a, \quad a \ne n,$$

$$\varphi_n = -2\gamma^n_n \wedge \varphi_n.$$

Put $\theta_a = \gamma^a_a - \gamma^{a+1}_{a+1}$, $\theta_n = 2\gamma^n_n$. θ_a is the complex connection form of (M,ds_a) written relative to φ_a.

Using the Maurer–Cartan structure equations and (‡) again we also obtain

$$d\gamma_1 = \varphi_1 \wedge \bar{\varphi}_1,$$

$$d\gamma_a = -\varphi_{a-1} \wedge \bar{\varphi}_{a-1} + \varphi_a \wedge \bar{\varphi}_a, \quad a \ne 1.$$

It now follows that

$$(*)\qquad \begin{aligned} d\theta_1 &= 2i(-2\Lambda_1 + \Lambda_2),\\ d\theta_a &= 2i(\Lambda_{a-1} - 2\Lambda_a + \Lambda_{a+1}),\\ d\theta_n &= 2i(\Lambda_{n-1} - 2\Lambda_n).\end{aligned}$$

Suppose M to be compact of genus *g*. Then the integration of (*) over M yields

$$2g - 2 - \#_a = \frac{1}{\pi}\int_M \Lambda_{a-1} - 2\Lambda_a + \Lambda_{a+1},$$

where $1 \leq a \leq n$, $\Lambda_0 = 0$, $\Lambda_{n+1} = \Lambda_{n-1}$, and $\#_a$ denotes the degree of the singular divisor of ds_a.

Notation. $d_a = \frac{1}{\pi}\int_M \Lambda_a$ $(= \frac{1}{\pi}\cdot(\text{the area of } M, ds_a))$.

It follows that

$$(**)\qquad 2g - 2 - \#_a = d_{a-1} - 2d_a + d_{a+1},$$

where $1 \leq a \leq n$, $d_0 = 0$, $d_{n+1} = d_{n-1}$. We call (**) the *symplectic Plücker formulae*. Observe that (d_a) determine $(\#_a)$, and vice versa.

The following is a corollary of the symplectic Plücker formulae. For a proof see [Y4] pp. 271–272.

Corollary. Let f: $M \to Sp(n)/T$ be a horizontal holomorphic *immersion* from a compact M. Then (i) for $n = 2$, K (the Gaussian curvature of the induced metric on M) $\geq 4/7$ implies that $K \equiv 4/7$, and (ii) for $n = 3$, $K \geq 2/11$ implies that $K \equiv 2/11$.

Chapter V. The Twistor Method

Let N be an oriented 2n–dimensional Riemannian manifold and also let $SO(N) \to N$ denote the $SO(2n)$–principal bundle of oriented orthonormal frames over N. The associated fibre bundle

$$SO(N) \times_{SO(2n)} SO(2n)/U(n) = SO(N)/U(n)$$

is called the *orthogonal twistor bundle* over N. The fibre at $x \in N$ parametrizes the set of all orientation–preserving orthogonal complex structures of the vector space T_xN. $\mathcal{T} = SO(N)/U(n)$ can be made into an almost complex manifold. In fact there are 2^γ, $\gamma = n(n-1)/2$, many natural almost complex structures on $\mathcal{T}$. (See §2 for the description.) And one attempts to study minimal surfaces in N in terms of complex curves in $\mathcal{T}$.

The most effective application of the twistor method occurs when the dimension of N is four. In this case any immersion $f\colon M^2 \to N$ lifts via the Gauss map to produce a map $T_f\colon M \to \mathcal{T}$. In §3 we prove that *f is minimal if and only if T_f is J_-–complex*, where as usual M^2 is given the induced complex structure. (The almost complex structure J_- on $\mathcal{T}$ will be defined in §2.)

For a survey of results on the twistor method the reader may consult [S2, E–S] and references cited therein. A good treatment of the case where N is a Riemannian symmetric space is given in [Br3].

§1. The Hermitian Symmetric Space SO(2n)/U(n)

Let $i\colon U(n) \to SO(2n)$ be the Lie group monomorphism induced by the identification

$$\mathbb{R}^{2n} \leftrightarrow \mathbb{C}^n, \quad (x^\alpha) \leftrightarrow (x^1 + ix^2, \cdots, x^{2n-1} + ix^{2n}).$$

We then have

(1) $$i(U(n)) = \{X \in SO(2n): {}^tXJ_nX = J_n\},$$

where $J_n = J_1 \oplus \cdots \oplus J_1 = \begin{bmatrix} J_1 & & \\ & \ddots & \\ & & J_1 \end{bmatrix}$, $J_1 = \begin{bmatrix} 0, & -1 \\ 1, & 0 \end{bmatrix}$.

Convention. We will identify U(n) with i(U(n)) and write U(n) instead of i(U(n)), $\mathfrak{u}(n)$ instead of $i_{*_e}(\mathfrak{u}(n))$, etc.

At the Lie algebra level we have

(2) $$i_{*_e}(\mathfrak{u}(n)) = \mathfrak{u}(n) = \{X \in \mathfrak{o}(2n): {}^tXJ_n + J_nX = 0\}.$$

Recall the vectors E_{ij}, F_{ij}, E'_{ij}, F'_{ij} $(1 \leq i < j \leq n) \in \mathfrak{o}(2n)$ and the root space decomposition (Chapter IV §3 (2))

(3) $$\mathfrak{o}(2n) = \mathfrak{t} \oplus \Sigma\, V_{ij} \oplus \Sigma\, V'_{ij},$$

where $V_{ij} = \mathbb{R}$–span $\{E_{ij}, F_{ij}\}$, $V'_{ij} = \mathbb{R}$–span $\{E'_{ij}, F'_{ij}\}$ and $\mathfrak{t}$ = the Lie algebra of $SO(2)^n \subset SO(2n)$. A routine computation then reveals that

(4) $$\mathfrak{u}(n) = \mathfrak{t} \oplus \Sigma\, V_{ij}.$$

It follows that

(5) $$\mathfrak{n} = \oplus\, \Sigma\, V'_{ij}$$

is the orthogonal complement to $\mathfrak{u}(n)$ in $\mathfrak{o}(2n)$ relative to the Killing form.

Via π_{*_e} (π: SO(2n) $\to$ SO(2n)/U(n)) $\mathfrak{n}$ is identified with the tangent space at the identity coset of SO(2n)/U(n).

(6) $$\mathfrak{n} \leftrightarrow T_0(SO(2n)/U(n)),\ 0 = U(n).$$

Let $\Omega = (\Omega^\alpha_\beta)$, $1 \leq \alpha,\beta \leq 2n$, denote the Maurer–Cartan form of SO(2n). The decompositions in (3–5) induce the decompositions

$$\Omega = \Omega_{\mathfrak{u}(n)} + \Omega_{\mathfrak{n}},\ \ \Omega_{\mathfrak{n}} = \Sigma\, \Omega_{V'_{ij}}.$$

We compute that

(7) $$\Omega_{V'_{ij}} = \tfrac{1}{2}\,[(\Omega^{2i-1}_{2j-1} - \Omega^{2i}_{2j}) \otimes E'_{ij} + (\Omega^{2i}_{2j-1} + \Omega^{2i-1}_{2j}) \otimes F'_{ij}] \text{ (no sum)}.$$

Put

(8) $$\Theta'^{ij} = \tfrac{1}{2}\,[(\Omega^{2i-1}_{2j-1} - \Omega^{2i}_{2j}) + \epsilon_{ij} i(\Omega^{2i}_{2j-1} + \Omega^{2i-1}_{2j})],$$

where $\epsilon_{ij} = \pm 1$.

Proposition. $SO(2n)/U(n)$ possesses exactly 2^{γ}, $\gamma = n(n-1)/2$, many invariant almost complex structures.

Proof. See [Y3] Chapter II §9. □

The invariant almost complex structures on $SO(2n)/U(n)$ can be described as follows: Pick $\{\epsilon_{ij} = +1 \text{ or } -1\colon 1 \leq i < j \leq n\}$ and let the pullbacks (by a local section of $SO(2n) \to SO(2n)/U(n)$) of (Θ'^{ij}) span type (1,0) forms on $SO(2n)/U(n)$. An invariant almost complex structure on $SO(2n)/U(n)$ corresponding to the choice $\{\epsilon_{ij}\}$ will be denoted by

(9) $$\oplus \Sigma\, \epsilon_{ij} J_1, \quad 1 \leq i < j \leq n.$$

Letting $\epsilon_{ij} = 1$ for every i and j we obtain an integrable invariant almost complex structure. $SO(2n)/U(n)$ has exactly one other integrable invariant almost complex structure which is the conjugate structure, corresponding to the choice $\epsilon_{ij} = -1$ for every i and j.

Example. Let $n = 2$. Then there are exactly two (both integrable, conjugate to each other) invariant complex structures on $SO(4)/U(2)$. Under $\epsilon_{12} J_1$ the pullback of

(10) $$\Theta'^{12} = \tfrac{1}{2}\,[(\Omega^1_3 - \Omega^2_4) + \epsilon_{12} i(\Omega^2_3 + \Omega^1_4)]$$

is of type (1,0).

Any invariant metric on $SO(2n)/U(n)$ is given by the pullback of the symmetric product

(11) $$c\cdot\Sigma\,\Theta'^{ij}\cdot\bar{\Theta}'^{ij},\quad c > 0.$$

Let < , > denote the standard inner product in $\mathbb{R}^{2n}$ and put

(12) $$\mathcal{J} = \{J \in \mathrm{Aut}^+(\mathbb{R}^{2n}) : J^2 = -I,\ \langle v, w\rangle = \langle Jv, Jw\rangle\},$$

where $\mathrm{Aut}^+(\mathbb{R}^{2n})$ denotes the set of orientation preserving automorphisms of $\mathbb{R}^{2n}$. $\mathcal{J}$ is the set of all orientation preserving orthogonal complex structures of the vector space $\mathbb{R}^{2n}$.

Let $A = (A_\alpha) = (A_1, \cdots, A_{2n}) \in SO(2n)$ and consider the assignment

(13) $$A \mapsto J_A \in \mathcal{J}$$

given by the formula

$$J_A(A_{2i-1}) = A_{2i} \text{ and } J_A(A_{2i}) = -A_{2i-1},\ 1 \le i \le n,$$

i.e., the matrix of J_A relative to the basis (A_α) is just J_n. This assignment induces a bijection

(14) $$SO(2n)/U(n) \overset{\cong}{\to} \mathcal{J}$$

as it is easily seen that $J_A = J_B$ if $B = A\cdot U$ for some $U \in U(n)$.

We have the following well–known symmetric space isomorphisms for n = 2, 3 (if n =1 then SO(2)/U(1) = a point):

(15) $$SO(4)/U(2) \cong SU(2)/U(1) \cong \mathbb{C}P^1,$$ $$SO(6)/U(3) \cong SU(4)/S(U(1)\times U(3)) \cong \mathbb{C}P^3.$$

§2. The Orthogonal Twistor Bundle

Let N denote a connected oriented 2n–dimensional Riemannian manifold and also let

(1) $$\pi\colon SO(N) \to N$$

denote the SO(2n)–principal bundle of oriented orthonormal frames over N.

The $\mathbb{R}^{2n}$–valued canonical form, denoted by $\Theta = (\Theta^\alpha)$, on SO(N) is given by

(2) $$\Theta(X) = e^{-1}(\pi_* X),\ X \in T_{(x,e)}SO(N)$$

where e is identified with a linear map $\mathbb{R}^{2n} \to T_xN$.

We have

(3) $$d\Theta^\alpha = -\Omega^\alpha_\beta \wedge \Theta^\beta,$$

where $\Omega = (\Omega^\alpha_\beta)$, $1 \le \alpha,\beta \le 2n$, is the $\mathfrak{o}(2n)$–valued Levi–Civita connection form on SO(N).

Definition. Put $\mathcal{T} = \{(x, J)\colon x \in N$, J is an orientation preserving orthogonal complex structure of the vector space $T_xN\}$. $\pi_2\colon \mathcal{T} \to N$, $(x, J) \mapsto x$, is called the *orthogonal twistor bundle over N*. ($\mathcal{T} \to N$ depends only on the conformal structure of N.)

Consider the projection

(4) $$\pi_1\colon SO(N) \to \mathcal{T},\ (x; e_1, \cdots, e_{2n}) \mapsto (x; J_e),$$

where $J_e\colon e_{2i-1} \mapsto e_{2i},\ e_{2i} \mapsto -e_{2i-1}$.

$$\begin{array}{ccc} SO(N) & \xrightarrow{\pi_1} & \mathcal{T} \\ & {}_{\pi}\searrow \quad \swarrow_{\pi_2} & \\ & N & \end{array}$$

Take $x \in N$ and fix an oriented orthonormal basis $\delta = (\delta_1, \cdots, \delta_{2n})$ of T_xN. For any $e = (e_\alpha) \in SO(N)_x = \pi^{-1}(x)$ we write $e_\alpha = e^\beta_\alpha \delta_\beta$ and obtain an identification

(5) $$SO(N)_x \cong SO(2n),\ e \mapsto (e^\beta_\alpha).$$

Likewise the isomorphism δ: $T_x N \to \mathbb{R}^{2n}$, $\delta_\alpha \mapsto (\partial/\partial x^\alpha)$, induces an identification

(6) $$\mathcal{T}_x \cong \mathcal{F} = SO(2n)/U(n).$$

Summing up the preceding discussion we have

(7) $$\mathcal{F} = SO(N) \times_{SO(2n)} SO(2n)/U(n) = SO(2n)/U(n),$$

and π_1: $SO(N) \to \mathcal{T}$ is a principal U(n)–fibration.

Recall from §1 the Lie algebra decomposition

(8) $$\mathfrak{o}(2n) = \mathfrak{u}(n) \oplus \mathfrak{n}, \quad \mathfrak{n} = \oplus \Sigma V'_{ij} \ (1 \le i < j \le n).$$

The $\mathfrak{o}(2n)$–valued connection form Ω on SO(N) decomposes accordingly,

(9) $$\Omega = \Omega_{\mathfrak{u}(n)} \oplus \Omega_{\mathfrak{n}}.$$

So,

(10) $$\Omega_{\mathfrak{n}} = \Sigma\, \Omega_{V'_{ij}} \ (1 \le i < j \le n),$$

where

$$\Omega_{V'_{ij}} = \tfrac{1}{2} [(\Omega^{2i-1}_{2j-1} - \Omega^{2i}_{2j}) \otimes E'_{ij} + (\Omega^{2i}_{2j-1} + \Omega^{2i-1}_{2j}) \otimes F'_{ij}] \text{ (no sum)}.$$

Put

(11) $$\Theta'^{ij} = \tfrac{1}{2} [(\Omega^{2i-1}_{2j-1} - \Omega^{2i}_{2j}) + \epsilon_{ij} i (\Omega^{2i}_{2j-1} + \Omega^{2i-1}_{2j})].$$

Define a symmetric bilinear form on SO(N) by

(12) $$B = {}^t\Theta \cdot \Theta + c \cdot \Sigma\, \Theta'^{ij} \cdot \bar{\Theta}'^{ij}, \quad c > 0.$$

We leave the proof of the following lemma as an exercise.

Lemma. i) $R_g^* B = B$, where $g \in U(n)$ and R_g denotes the right multiplication by g on SO(N); ii) B(v, w) = 0 if one of v, w is a vertical vector of the fibration $SO(N) \to \mathcal{T}$.

As a consequence of the above lemma there exists a unique Riemannian metric, denoted by ds^2_T, on $\mathcal{T}$ such that

(13) $$\pi_1^* \, ds_T^2 = B.$$

Remark. $(SO(N), B) \to (\mathcal{T}, ds_T^2)$ is a Riemannian submersion with totally geodesic fibres.

Let $(x, J) \in \mathcal{T}$ and $V_{(x,J)}$ denote the subspace of $T_{(x,J)}\mathcal{T}$ tangential to the fibre $\pi_2^{-1}(x)$. We also let $H_{(x,J)} \subset T_{(x,J)}\mathcal{T}$ denote the orthogonal complement to $V_{(x,J)}$ with respect to the metric ds_T^2. ($H_{(x,J)}$ is well–defined independent of the choice c in (12).)

(14) $$T_{(x,J)}\mathcal{T} = H_{(x,J)} \oplus V_{(x,J)}.$$

The distribution defined by $H_{(x,J)}$, $(x,J) \in \mathcal{T}$, will be called the *horizontal distribution* of $\mathcal{T} \to N$.

π_{2*} gives an isomorphism $H_{(x,J)} \cong T_xN$. Now J is a complex structure on T_xN, hence via the above isomorphism J defines a complex structure on $H_{(x,J)}$. On the other hand $\pi_2^{-1}(x) \cong SO(2n)/U(n)$ and an invariant almost complex structure on $SO(2n)/U(n)$ gives rise to a complex structure on $V_{(x,J)}$. More precisely, pick an oriented orthonormal frame at $x \in N$ and use it to identify T_xN with $\mathbb{R}^{2n}$. This identification induces another identification $\pi_2^{-1}(x) \leftrightarrow \mathcal{T} = SO(2n)/U(n)$. Thus $V_{(x,J)}$ is identified with the tangent space at some point of $SO(2n)/U(n)$ and translating this tangent space to the identity coset of $SO(2n)/U(n)$ we obtain the identification

$$V_{(x,J)} \leftrightarrow \mathfrak{n} = \mathfrak{u}(n)^{\perp} \subset \mathfrak{o}(2n).$$

An invariant almost complex structure on $SO(2n)/U(n)$ by restriction defines a complex structure on the vector space $\mathfrak{n}$, hence one on $V_{(x,J)}$. Taking the direct sum of this "vertical" complex structure with the canonical complex structure defined on $H_{(x,J)}$ above at every $(x, J) \in \mathcal{T}$ we obtain an almost complex structure on $\mathcal{T}$. To put it another way, we obtain an almost complex structure on $\mathcal{T}$ by decreeing that the 1–forms on SO(N) given by

$$(15) \qquad \Phi^i = \Theta^{2i-1} + i\Theta^{2i} \ (1 \le i \le n), \quad \Theta'^{ij} \ (1 \le i < j \le n)$$

pull back (by a local section) to give type (1,0) forms on $\mathcal{T}$. This almost complex structure will be denoted by

$$(16) \qquad J_H \oplus \Sigma\, \epsilon_{ij} J_1 .$$

In particular we have

Proposition. There are 2^γ, $\gamma = n(n-1)/2$, many *natural* almost complex structures on $\mathcal{T} \to N$, where dim $N = 2n$.

Example. Let $N = S^4$. Then $SO(N) = SO(5)$ and

$$\mathcal{T} = SO(5)/U(2) \cong Sp(2)/U(1){\times}Sp(1) \cong \mathbb{C}P^3.$$

The complex structure on $\mathbb{C}P^3$ corresponds to $J_H \oplus \Sigma\, \epsilon_{12} J_1$ with $\epsilon_{12} = 1$.

Notation. $J_+ = J_H \oplus \Sigma\, \epsilon_{ij} J_1$ with every $\epsilon_{ij} = 1$; $J_- = J_H \oplus \Sigma\, \epsilon_{ij} J_1$ with every $\epsilon_{ij} = -1$.

Remark. The almost complex structures J_+ and J_- have been studied extensively by various authors, e.g., see [Br3, Bu, E–S, J–R2, Raw, S1, S2]. However, the remaining almost complex structures on $\mathcal{T}$, to the author's knowledge, have not been explored.

Exercise. With the possible exception of J_+ the above almost complex structures on $\mathcal{T}$ are never integrable. (Exterior differentiate the forms Φ^i, Θ'^{ij}.)

§3. Applications: Isotropic Surfaces and Minimal Surfaces

Let N be an 2n–dimensional connected oriented Riemannian manifold and also let M be a Riemann surface. We consider a conformal immersion

(1) $$f: M \to N.$$

Index Convention. $1 \le \alpha,\beta,\gamma \le 2n$; $1 \le i,j,k \le 2$; $3 \le a,b,c \le 2n$.

Definition. A Darboux frame along f is a local oriented orthonormal frame $e = (e_\alpha)$ in N such that $(e_i \circ f)$ is a local oriented orthonormal frame in $(M, f^* ds_N^2)$ and $(e_a \circ f)$ are normal to M.

In particular e is a local section of $SO(N) \to N$ and its dual coframe is $(e^* \Theta^\alpha)$, where $\Theta = (\Theta^\alpha)$ is the canonical form on SO(N). Moreover, for a Darboux frame along f we have

(2) $$f^* e^* \Theta^a = 0.$$

We will write just Θ^α instead of $e^* \Theta^\alpha$ suppressing e^*. Likewise we write Ω^α_β instead of $e^* \Omega^\alpha_\beta$, where $\Omega = (\Omega^\alpha_\beta)$ is the Levi–Civita connection form on SO(N). In keeping with this convention we also confuse $e_\alpha \circ f$ with e_α and let the context dictate the proper meaning.

Notation. $f^* \Theta^\alpha = \theta^\alpha$, $f^* \Omega^\alpha_\beta = \omega^\alpha_\beta$.

The induced metric $f^* ds_N^2$ is then given by

(3) $$\varphi \cdot \bar{\varphi}, \quad \text{where } \varphi = \theta^1 + i\theta^2.$$

The conformality of f is manifested by the fact that φ is a local type (1,0) form on M.

Exterior differentiation of both sides of the equations in (2) leads to

(4) $$\omega^a_i = h^a_{ij} \theta^j,$$

for some local functions $h^a_{ij} = h^a_{ji}$ on M.

The second fundamental tensor of f is

(5) $$II = h^a_{ij}\, \theta^i \cdot \theta^j \otimes e_a.$$

The mean curvature vector of f is

(6) $$H = H^a e_a, \text{ where } H^a = \frac{1}{2}(h^a_{11} + h^a_{22}).$$

f is minimal iff $H \equiv 0$.

Put

(7) $$L^a = \frac{1}{2}\cdot(h^a_{11} - h^a_{22}) - ih^a_{12}.$$

If $\tilde{e}$ denotes another Darboux frame along f then on their common domain e and $\tilde{e}$ are related by

(8) $$\tilde{e} = e\cdot k,$$

where k is $SO(2)\times SO(2n-2)$–valued. We put

(9) $$k = (A, B),\quad A = \begin{pmatrix}\cos t, -\sin t\\ \sin t, \cos t\end{pmatrix},\ B \in SO(2n-2).$$

Define tilded quantities $\tilde{\varphi}$, $\tilde{h}^a$, $\tilde{H}^a$, $\tilde{L}^a$ using $\tilde{e}$ (e.g., $\tilde{\varphi} = \tilde{\theta}^1 + i\tilde{\theta}^2$, where $\tilde{\theta}^i = f^*\tilde{e}^*\Theta^i$). We then obtain the following set of transformation rules whose verifications are routine and left to the reader.

(10) $$\tilde{\varphi} = e^{-it}\varphi,$$

(11) $$\tilde{h}^a = ({}^tB)^a_b\, {}^tAh^bA,$$

(12) $$\tilde{H}^a = ({}^tB)^a_b\, H^b,$$

(13) $$\tilde{L}^a = e^{2it}\, ({}^tB)^a_b\, L^b.$$

From these transformation rules we obtain at once

Proposition. The symmetric type (4,0) form given by

(14) $$\Lambda = (\Sigma\, L^a\cdot L^a)(\varphi)^4$$

is globally defined on M.

Definition. A conformal immersion $f: M \to N$ is said to be *isotropic* (sometimes called real–isotropic) if $\Lambda \equiv 0$.

Example. Consider a conformal minimal immersion f: $S^2 = \mathbb{C}P^1 \to S^{2n}$. Then Chapter II §2 (11) shows that f is also isotropic. Indeed using Theorem 2 of Chapter II §2 we had a (super–)horizontal holomorphic lifting

$$\Phi_f\colon \mathbb{C}P^1 \to SO(2n+1)/SO(2)^n.$$

The inclusion $SO(2)^n = U(1)^n \subset U(n)$ induces a projection

$$\pi\colon SO(2n+1)/SO(2)^n \to SO(2n+1)/U(n).$$

Now $SO(2n+1) = SO(S^{2n})$ and $SO(2n+1)/U(n) \cong \mathcal{T}$, the twistor bundle over S^{2n}. We thus obtain a lifting of f

$$\pi\circ\Phi_f\colon \mathbb{C}P^1 \to SO(2n+1)/SO(2)^n \to \mathcal{T}$$

and it is easily verified that $\pi\circ\Phi_f$ is horizontal relative to the horizontal distribution of $\mathcal{T} \to S^{2n}$. We leave it to the reader to check that this map is in fact complex with respect to any one of the 2^γ almost complex structures on $\mathcal{T}$ defined in §2. (Notice that for a horizontal map h: $M \to \mathcal{T}$, h is complex with respect to one of the almost complex structures $J_H \oplus \Sigma\, \epsilon_{ij} J_1$ if and only if it is complex with respect to the remaining almost complex structures.)

The following result is proved in [S2] §2.

Theorem. i) The projection to N of a J_+–complex curve in $\mathcal{T}$ is isotropic, possibly with branch points; ii) The projection to N of a J_-–complex curve in $\mathcal{T}$ is a generalized conformal minimal immersion.

A sort of converse to the above theorem holds when the dimension of the target manifold N is four. So we consider a conformal immersion

(15) $$f\colon M \to N,$$

where M is a Riemann surface and N is a conencted oriented 4–dimensional Riemannian manifold.

Definition. The *twistor lift* of f is the map

$$T_f\colon M \to \mathcal{T} = SO(N)/U(2)$$

given by

$$x \mapsto (x, J_x);\quad J_x\colon e_1(x) \mapsto e_2(x),\ e_3(x) \mapsto e_4(x),$$

where (e_α) is a Darboux frame along f.

Let $G_2(TN)$ denote the Grassmann bundle of oriented tangent two–planes of N. We have an identification

$$G_2(TN) \leftrightarrow SO(N)/SO(2)^2$$

and the inclusion $SO(2)^2 = U(1)^2 \subset U(2)$ induces a projection

$$\pi\colon G_2(TN) \to \mathcal{T}.$$

Let $\Psi_f\colon M \to G_2(TN)$ denote the Gauss map. We then have

(16) $$T_f = \pi \circ \Psi_f.$$

The following theorem was first proved in [E–S].

Theorem. Let $f\colon M \to N^4$ be a conformal immersion from a Riemann surface. Then i) f is isotropic if and only if T_f is J_+–complex; ii) f is minimal if and only if T_f is J_-–complex.

Proof. We will prove ii), i) being similar. Let e be a Darboux frame along f. We have

$$f^*e^*\Theta'^{12} = \frac{1}{2}\,[(\omega_3^1 - \omega_4^2) + \epsilon_{12} i(\omega_4^1 + \omega_3^2)],$$

$$f^*e^*(\Theta^1 + i\Theta^2) = \varphi \in \text{type } (1,0).$$

Now $e \circ f = u \circ T_f$ for some local section u of $SO(N) \to \mathcal{T}$ and $u^*\Theta'^{12}$ with $\epsilon_{12} = -1$ is a type (1,0) form in $(\mathcal{T}, J_-)$. It follows that T_f is J_-–complex if and only if $f^*e^*\Theta'^{12}$ with $\epsilon_{12} = -1$ is of type (1,0) on M. Now

$$-2f^*e^*\Theta'^{12} = (\omega_1^3 - i\omega_2^3) - i(\omega_1^4 - i\omega_2^4).$$

From this we compute that

$$-2f^*e^*\Theta'^{12} = (H^3 - iH^4)\bar{\varphi} + (L^3 - iL^4)\varphi.$$

So $f^*e^*\Theta'^{12}$ (with $\epsilon_{12} = -1$) is of type (1,0) iff $H^3 - iH^4 \equiv 0$ iff $H^3 = H^4 \equiv 0$ iff f is minimal. □

§4. Self–Duality in Riemannian Four–Manifolds

Let N be an oriented Riemannian 2n dimensional manifold, and also let $\mathcal{T} \to N$ denote the orthogonal twistor bundle. Although the almost complex structure J_+ on $\mathcal{T}$ is not directly used to study minimal surfaces in N, it is the only structure that is possibly integrable. For this reason J_+ plays an important role in Mathematical Physics, especially in Yang–Mills theory. In this section we give a cursory look at the theory of instantons (which are certain solutions to the Yang–Mills equation) as developed by Penrose, Atiyah and others. Our main reference for this section is the article [A–H–S]. In particular, proofs of, and further references to, the results in this section can be found there.

Let N be an oriented Riemannian four–manifold, and also let $\Lambda^2(N)$ denote the space of smooth 2–forms on N. The Hodge star operator gives a $C^\infty(N)$–linear map

$$*: \Lambda^2(N) \to \Lambda^2(N),$$
$$*(\theta^{\sigma(1)} \wedge \theta^{\sigma(2)}) = \mathrm{sign}(\sigma)\ \theta^{\sigma(3)} \wedge \theta^{\sigma(4)},$$

where (θ^i) is an oriented orthonormal coframe and σ is a permutation.

Note that $(*)^2$ = the identity. Hence we obtain a decomposition

$$(\dagger) \qquad \Lambda^2(N) = \Lambda^2_+(N) \oplus \Lambda^2_-(N),$$

where $\Lambda^2_\pm(N)$ are the ±1 eigenspaces of *.

The Levi–Civita connection forms, $\omega = (\omega^i_j)$, satisfy

$$d\theta^i = -\omega^i_j \wedge \theta^j.$$

The curvature forms, $\chi = (\chi^i_j)$, are given by

$$\chi^i_j = d\omega^i_j + \omega^i_k \wedge \omega^k_j.$$

The curvature form χ can be thought of as a self–adjoint transformation

$$\chi\colon \Lambda^2(N) \to \Lambda^2(N), \quad \chi(\theta^i \wedge \theta^j) = \chi^i_j.$$

Thus χ decomposes according to the decomposition given in (†). The "matrix" of χ then looks like

$$\chi = \begin{bmatrix} A, & B \\ {}^tB, & C \end{bmatrix},$$

where $A \in \mathrm{End}(\Lambda^2_+(N))$, $C \in \mathrm{End}(\Lambda^2_-(N))$, $B \in \mathrm{Hom}(\Lambda^2_-(N), \Lambda^2_+(N))$. N is *Einstein* if and only if $B \equiv 0$. Also $\mathrm{trace}(A) = \mathrm{trace}(C) = \frac{1}{4}\cdot$(scalar curvature). Put

$$W_+ = A - \tfrac{1}{3}\mathrm{trace}(A), \quad W_- = C - \tfrac{1}{3}\mathrm{trace}(C).$$

Then $W = W_+ + W_-$ is conformally invariant, and is called the *Weyl tensor.* It is not hard to show that N is conformally flat iff $W \equiv 0$.

Definition. The Riemannian four–manifold N is said to be self–dual if $W_- = 0$. N is said to be anti–self–dual if $W_+ = 0$.

As * and W are conformal invariants the notion of self–duality depends only on the conformal structure of N.

Example. The Euclidean four–sphere S^4 is self–dual. In fact, the underlying conformal structure on the Euclidean four–sphere is the unique conformal structure on the topological four–sphere that makes it self–dual.

Let $\mathcal{T} \to N^4$ denote the orthogonal twistor bundle. We have

Proposition ([A–H–S]). J_+ on $\mathcal{T}$ is integrable if and only if N is anti–self–dual.

Let $\hat{\mathcal{T}} \to N$ denote the twistor bundle whose fibre at $x \in N$ consists of all orientation *reversing* orthogonal complex structures of T_xN. We then have:

J_+ on $\hat{\mathcal{T}}$ is integrable if and only if N is self–dual.

Let $P \to N$ be a principal bundle with a compact structure group G. A connection on P is defined by a $\mathfrak{g}$–valued 1–form ω_P. Let ad(P) denote the vector bundle associated to P via the adjoint representation of G on its Lie algebra $\mathfrak{g}$. The curvature of ω_P is given by the ad(P)–valued 2–form

$$\chi_P = d\omega_P + \omega_P \wedge \omega_P.$$

Physicists call χ_P the *gauge field.*

Definition. A connection ω_P is said to be self–dual (respectively, anti–self–dual) if its curvature χ_P satisfies $^*\chi_P = \chi_P$ (respectively, $^*\chi_P = -\chi_P$).

Using the Riemannian metric on N together with the Cartan–Killing form on $\mathfrak{g}$ we can define a square norm $|\cdot|_x^2$ for the curvature χ_P at each $x \in N$. Relative to an orthonormal basis of T_xN and an orthonormal basis for $\mathfrak{g}$ this norm is simply the sum of the squares of the absolute value of the components of χ_P at $x \in N$.

Definition. The Yang–Mills functional of N is given by

$$\chi_P \mapsto \int_N |\chi_P|_x^2 \, dx,$$

where dx is the volume element of N.

The Yang–Mills functional takes on finite values when N is compact.

A fundamental observation is that: *A (anti–) self–dual connection gives an absolute minimum for the Yang–Mills functional.* When G = SU(2) and N = S^4 a self–dual connection is often called an *instanton.* The totality of instantons corresponds to a certain class of holomorphic bundles (namely, holomorphic bundles which are trivial on the fibres of $\hat{\mathcal{T}} \to S^4$, and with a real structure inducing positive definite Hermitian inner products on the fibres) over $\hat{\mathcal{T}} = \mathbb{C}P^3$. For a full description of this correspondence see [A–H–S].

BIBLIOGRAPHY

[A] L.V. Ahlfors, *Complex Analysis*, 2nd ed., McGraw–Hill, New York, 1966.

[A–S] L.V. Ahlfors and L. Sario, *Riemann Surfaces*, Princeton University Press, New Jersey, 1960.

[Al1] F.J. Almgren, Some interior regularity theorems for minimal surfaces and an extension of Bernstein's theorem, Ann. of Math. 84 (1966), 277–292.

[Al2] F.J. Almgren, *Plateau's Problem*, Benjamin, New York, 1966.

[A–H–S] M.F. Atiyah, H.J. Hitchin and I.M. Singer, Self–duality in four dimensional Riemannian geometry, Proc. Royal Soc. London Ser. A 362 (1978), 425–461.

[B–C] J.L. Barbosa and A.G. Colares, *Minimal Surfaces in* $\mathbb{R}^3$, Lecture Notes in Math., No. 1195, Springer–Verlag, New York, 1986.

[B–D1] J.L. Barbosa and M. Do Carmo, On the size of a stable minimal surface in $\mathbb{R}^3$, Amer. J. Math. 98 (1976), 515–528.

[B–D2] J.L. Barbosa and M. Do Carmo, A necessary condition for a metric in M^n to be minimally immersed in R^{n+1}, An. Acad. Bras. Cienc. 50 (1978), 445–454.

[B–D3] J.L. Barbosa and M. Do Carmo, Stability of Minimal surfaces and eigenvalues of the Laplacian, Math. Zeitschrift 173 (1980), 13–28.

[B] S. Bernstein, Sur un theoreme de geometrie et ses applications aux equations aux derives partielles du type elliptique, Comm. Soc. Math. Kharkov 15 (1915–1917), 38–45.

[Bo] E. Bombieri, Recent progress in the theory of minimal surfaces, L'Enseignement Math. 25 (1975), 1–8.

[B–D–G] E. Bombieri, E. DeGiorgi and E. Guisti, Minimal cones and the Bernstein problem, Invent. Math. 7 (1969), 243–268.

[Br1] R.L. Bryant, Conformal and minimal immersions of compact surfaces into the 4–sphere, J. Diff. Geom. 17 (1982), 455–473.

[Br2] R.L. Bryant, Minimal surfaces of constant curvature in S^n, Trans. Amer. Math. Soc. 290 (1985), 259–271.

[Br3] R.L. Bryant, Lie groups and twistor spaces, Duke Math. J. 52 (1985), 223–262.

[Bu] F. Burstall, Twistor methods for harmonic maps, pp. 55–96 in *Differential Geometry – Proceedings of the Nordic Summer School*, Lecture Notes in Math. Vol. 1263, Springer–Verlag, 1987.

[C1] E. Calabi, An extension of E. Hopf's Maximum Principle with an application to Riemannian geometry, Mich. Math. J. 25 (1958), 45–56.

[C2] E. Calabi, Minimal immersions of surfaces in Euclidean spheres, J. Diff. Geom. 1 (1967), 111–125.

[C3] E. Calabi, Quelques applications de l'analyse complexe aux surfaces d'aire minime, pp. 59–81 in *Topics in Complex Manifolds*, H. Rossi ed., Univ. of Montreal, Montreal, 1968.

[C] E. Cartan, *Théorie des Groupes Finis et Continus et la Géométrie Différentielle traitées par la Méthode du Repère Mobile*, Gauthier–Villars, Paris, 1937

[Ch] I. Chavel, *Eigenvalues in Riemannian Geometry*, Academic Press, New York, 1984.

[Che1] C.C. Chen, Complete minimal surfaces with total curvature -2π, Boletim da Soc Mat. Brasil. 10 (1979), 71–76.

[Che2] C.C. Chen, A characterization of the catenoid, An. Acad. Brasil Cienc. 51 (1979), 1–3.

[Che3] C.C. Chen, Elliptic functions and non–existence of complete minimal surfaces of certain type, Proc. Amer. Math. Soc. 79 (1980), 289–293.

[Che4] C.C. Chen, The generalized curvature ellipses and minimal surfaces, Bull. Inst. Math. Acad. Sinica 11 (1983), 329–336.

[C–G] C.C. Chen and F. Gackstatter, Elliptische und hyperelliptische Funktionen und vollständige Minimalflächen vom Enneperschen Typ, Math. Annalen, 259 (1982), 359–370.

[C–Y] S.–Y. Cheng and S.T. Yau, Differential equations on Riemannian manifolds and their geometric applications, Comm. Pure Appl. Math. 28 (1975), 333–354.

[Cher1] S.S. Chern, An elementary proof of the existence of isothermal parameters on a surface, Proceed. of Amer. Math. Soc., 6 (1955), 771–782.

[Cher2] S.S. Chern, Minimal surfaces in an Euclidean space of N dimensions, pp. 187–198 in *Differential and Combinational Topology*, A symposium in honor of Marston Morse, Princeton University Press, Princeton, 1965.

[Cher3] S.S. Chern, Simple proofs of two theorems on minimal surfaces, L'Enseignement Math. XV (1969), 53–61; or *Selected Papers*, 361–369.

[Cher4] S.S. Chern, On the minimal immersions of the two–sphere in a space of constant curvature, pp. 27–40 in *Problems in Analysis*, Princeton Univ. Press, Princeton, 1970.

[C–D–K] S.S. Chern, M. Do Carmo and S. Kobayashi, Minimal submanifolds of a sphere with second fundamental form of constant length, *Functional Analysis and Related Fields*, F.E. Browder, ed., Springer–Verlag, Berlin, 1970.

[C–O1] S.S. Chern and R. Osserman, Complete minimal surfaces in Euclidean n–space, J. d'Analyse Math. 19 (1967), 15–34.

[C–O2] S.S. Chern and R. Osserman, Remarks on the Riemannian metric of a minimal submanifold, *Geometry Symposium, Utrecht 1980*, Springer Lecture Notes No. 894, Springer–Verlag, New York, 1981, 49–88.

[C–W1] S.S. Chern and J. Wolfson, Minimal surfaces by moving frames, Amer. J. Math. 105 (1983), 59–83.

[C–W2] S.S. Chern and J. Wolfson, Harmonic maps of the two–sphere into a complex Grassman manifold II, Ann. of Math. 125 (1987),301–335.

[C–S] H.I. Choi and R. Schoen, The space of minimal embeddings of a surface into a three–dimensional manifold of positive Ricci curvature, preprint.

[Cho–W] H.I. Choi and A. Wang, A first eigenvalue estimate for minimal hypersurfaces, J. Diff. Geom. 18 (1983), 559–562.

[Co] C.J. Costa, Example of a complete minimal immersion in R^3 of genus one and three embedded ends, Bol. Soc. Bras. Mat. 15 (1984), 47–54.

[Cou] R. Courant, *Dirichlet's Principle, Conformal Mapping, and Minimal Surfaces*, Interscience Publishers, New York, 1950.

[D–Z] A.M. Din and W.J. Zakrzewski, General classical solutions in the CP^n model, Nucl. Phys. B 174 (1980), 397–406.

[D–D] M. Do Carmo and M. Dajczer, An infinite family of simply connected minimal surfaces in H^3, Trans. Amer. Math. Soc. 277 (1983), 685–709.

[D–P] M. Do Carmo and C.K. Peng, Stable minimal surfaces in $\mathbb{R}^3$ are planes, Bull. Amer. Math. Soc. 1 (1979), 903–906.

[D–W1] M. Do Carmo and N. Wallach, Representations of compact groups and minimal immersions into spheres, J. Diff. Geom. 4 (1970), 91–104.

[D–W2] M. Do Carmo and N. Wallach, Minimal immersions of spheres into spheres, Ann. of Math. 93 (1971), 43–62.

[D] J. Douglas, Solution of the problem of Plateau, Trans. Amer. Math. Soc. 33 (1931), 263–321.

[E–S] J. Eells and S. Salamon, Twistorial constructions of harmonic maps of surfaces into four–manifolds, Ann. Scuola Norm. Sup. Pisa 12 (1985), 589–640.

[E–W] J. Eells and J.C. Wood, Harmonic maps from surfaces to complex projective spaces, Advances in Math., 49 (1983), 217–263.

[F–K] H.M. Farkas and I. Kra, *Riemann Surfaces*, Springer–Verlag, New York, 1980.

[F] H. Federer, *Geometric Measure Theory*, Springer–Verlag, New York, 1969.

[Fi] R. Finn, On a class of conformal metrics, with application to differential geometry in the large, Comment. Math. Helv., 40 (1965), 1–30.

[F–C] D. Fischer–Colbrie and R. Schoen, The structure of complete stable minimal surfaces in 3–manifolds of non–negative scalar curvature, Comm. Pure Appl. Math. 33 (1980), 199–211.

[Fl] W.H. Fleming, On the oriented Plateau problem, Rend. Circolo Mat. Palermo 9 (1962), 69–90.

[Fu1] H. Fujimotto, On the Gauss map of a complete minimal surface in R^n, J. Math. Soc. Japan 35 (1983), 279–288.

[Fu2] H. Fujimoto, Value distribution of the Gauss maps of complete minimal surfaces in R^n, J. Math. Soc. Japan 35 (1983), 663–681.

[Fu3] H. Fujimoto, On the number of exceptional values of the Gauss maps of minimal surfaces, J. Math. Soc. japan 40 (1988), 235–248.

[G–K] F. Gackstatter and F. Kunert, Konstruktion vollständiger Minimalflächen von endlicher Gesamtkrümmung, Arc. Rat. Mech. Anal., 65 (1977), 289–297.

[Gl] J.F. Glazebrook, The construction of a class of harmonic maps to quaternionic projective spaces, J. London Math. 30 (1984), 151–159.

[G–O–R] R.D. Gulliver, R. Osserman and H.L. Royden, A theory of branched immersions of surfaces, Amer. J. Math. 95 (1973), 750–812.

[G–T] D. Gilbarg and N.S. Trudinger, *Elliptic Partial Differential Equations of Second Order*, Springer–Verlag, New York, 1977.

[G] E. Giusti, *Minimal Surfaces and Functions of Bounded Variation*, Birkhäuser, Boston, 1984.

[Gr] P. Griffiths, On Cartan's method of Lie groups and moving frames as applied to uniqueness and existence questions in differential geometry, Duke Math. J., 41 (1974), 775–814.

[G–H] P. Griffiths and J. Harris, *Principles of Algebraic Geometry*, Wiley, New York, 1978.

[G–S] R.D. Gulliver and J. Spruck, On embedded minimal surfaces, Ann.of Math. 103 (1976), 331–347.

[H] J. Hass, Complete area minimizing minimal surfaces which are not totally geodesic, Pacific J. Math. 111 (1984), 35–38.

[H–H] E. Heinz and S. Hildebrandt, Some remarks on minimal surfaces in Riemannian manifolds, Comm. Pure and Appl. Math 23 (1970), 371–377.

[Ho] D.A. Hoffman, Lower bounds on the first eigenvalue of the Laplacian of Riemannian manifolds, pp. 61–72 in *Minimal Submanifolds and Geodesics*, M. Obata, ed., Elsevier/North–Holland, New York, 1979.

[H–M1] D.A. Hoffman and W. Meeks, III., Complete embedded minimal surfaces of finite total curvature, Bull. Amer. Math. Soc. 12 (1985), 134–136.

[H–M2] D.A. Hoffman and W. Meeks, III., A complete embedded minimal surface in R^3 with genus one and three ends, J. of Diff. Geom. 21 (1985), 109–127.

[H–O1] D.A. Hoffman and R. Osserman, The geometry of the generalized Gauss map, Mem. Amer. Math. Soc. Vol. 28, No. 236, 1980.

[H–O2] D.A. Hoffman and R. Osserman, The area of the generalized Gaussian image and the stability of minimal surfaces in S^n and R^n, Math. Ann. 260 (1982), 437–452.

[Hop1] H. Hopf, Uber Flachen mit einer Relation Zwischen den Haupt–krumungen, Math. Nachr. 4 (1951), 232–249.

[Hop2] H. Hopf, *Lectures on Differential Geometry in the Large*, Springer Lectures Notes No. 1000, Springer–Verlag, New York, 1983.

[Hs1] W.–Y. Hsiang, On compact homogeneous minimal submanifolds, Proc. Nat. Acad. Sci. U.S.A., 56 (1966), 5–6.

[Hs2] W.–Y. Hsiang, Generalized rotational hypersurfaces of constant mean curvature in the euclidean spaces I, J. Diff. Geom. 17 (1982), 337–356.

[Hs3] W.–Y. Hsiang, Minimal cones and the spherical Bernstein problem I, Annals of Math. 118 (1983), 61–73.

[Hs4] W.–Y. Hsiang, Minimal cones and the spherical Bernstein problem II, Invent. Math. 74 (1983), 351–369.

[H–L] W.–Y. Hsiang and H.B. Lawson, Minimal submanifolds of low cohomogeneity, J. Diff. Geom. 5 (1971), 1–38.

[Hu] A. Huber, On subharmonic functions and differential geometry in the large, Comment. Math. Helv. 32 (1957), 13–72.

[J] G.R. Jensen, *Higher Order Contact of Submanifolds of Homogeneous Spaces*, Lecture Notes in Math., Vol. 610, Springer–Verlag, New York, 1977.

[J–R1] G.R. Jensen and M. Rigoli, Minimal surfaces in spheres, Special Volume, Rend. Sem. Mat. Univers. Politecn. Torino (1983), 75–98.

[J–R2] G.R. Jensen and M. Rigoli, Twistors and Gauss lifts of surfaces in four–manifolds, preprint.

[J–R–Y] G.R. Jensen, M. Rigoli and K. Yang, Holomorphic curves in the complex quadric, Bull. Austral. Math. Soc. 35 (1978), 125–147.

[J–M] L.P. Jorge and W. Meeks, III, The topology of complete minimal surfaces of finite total Gaussian curvature, Topology 22 (1983), 203–221.

[J–X1] L.P. Jorge and F. Xavier, On the existence of complete bounded minimal surfaces, Bol. Soc. Brasil. Mat. 10 (1979).

[J–X2] L.P. Jorge and F. Xavier, A complete minimal surface in R^3 between two parallel planes, Ann. of Math. 112 (1980), 203–206.

[K–P–S] H. Karcher, U. Pinkall and I. Sterling, New minimal surfaces in S^3, J. Diff. Geom. 28 (1988), 169–186.

[K] L. Karp, Subharmonic functions on real and complex manifolds. Math. Zeitschrift 179 (1982), 535–554.

[Ka] S. Kawai, A theorem of Bernstein type for minimal surfaces in R^4, Tohoku Math. J. 36 (1984), 377–384.

[Ke1] K. Kenmotsu, On compact minimal surfaces with nonnegative Gaussian curvature in a space of constant curvature I, II, Tohoku Math. J. 25 (1973), 469–479; 27 (1975), 291–301.

[Ke2] K. Kenmotsu, On a parametrization of minimal immersions of R^2 into S^5, Tohoku Math J. 27 (1975), 83–90.

[Ke3] K. Kenmotsu, On minimal immersions of R^2 into S^n, J. Math. Soc. Japan 28 (1976), 182–191.

[Ke4] K. Kenmotsu, Minimal surfaces with constant curvature in 4–dimensional space forms, Proc. Amer. Math. Soc. 89 (1983), 133–138.

[K–S] T. Klotz and L. Sario, Existence of complete minimal surfaces of arbitrary connectivity and genus, Proc. Nat. Acad. Sci. 54 (1965), 42–44.

[Ko] M. Koiso, On the stability of minimal surfaces in R^3, J. Math. Soc. Japan 36 (1984), 523–541.

[L–O] H.B. Lawson, Jr. and R. Osserman, Non–existence, non–uniqueness and irregularity of solutions to the minimal surface system, Acta Math. 139 (1977), 1–17.

[L1] H.B. Lawson, Jr., Complete minimal surfaces in S^3, Ann. of Math. 92 (1970), 335–374.

[L2] H.B. Lawson, Jr., The global behavior of minimal surfaces in S^n, Ann. Math. 92 (1970), 224–237.

[L3] H.B. Lawson, Jr., *Lectures on Minimal Submanifolds*, Vol. I, Publish or Perish, Berkeley, 1980.

[L4] H.B. Lawson, Jr., Some intrinsic characterizations of minimal surfaces, J. d'Anal. Math. 24 (1971), 151–161.

[L5] H.B. Lawson, Jr., *Minimal Varieties in Real and Complex Geometry*, Univ. of Montreal Press, Montreal, 1973.

[Le] H. Lewy, On the boundary behavior of minimal surfaces, Proc. Nat'l Acad. Sci. 37 (1951), 103–110.

[M1] W. Meeks, III, Uniqueness theorems for minimal surfaces, Illinois J. Math. 25 (1981), 319–335.

[M2] W. Meeks, III, The topological uniqueness of minimal surfaces, Topology 20 (1981), 389–410.

[M–Y] W. Meeks, III and S.T. Yau, Topology of three dimensional manifolds and the embedding problems in minimal surface theory, Ann. of Math. 112 (1980), 441–485.

[Mi] M.J. Micallef, Stable minimal surfaces in Euclidean space, J. Diff. Geom. 19 (1984), 57–84.

[M–O] X. Mo and R. Osserman, On the Gauss map and total curvature of complete minimal surfaces and an extension of Fujimoto's theorem, preprint.

[Mo] J.D. Moore, On stability of minimal spheres and a two–dimensional version of synge's theorem, Arch. Math. (Basel) 4444 (1985), 278–281.

[Mor1] F. Morgan, Almost every curve in R^3 bounds a unique area minimizing surface, Inventiones Math. 45 (1978), 253–297.

[Mor2] F. Morgan, A smooth curve in R^4 bounding a continuum of area minimizing surfaces, Duke Math. J. 43 (1976), 867–870.

[Mori1] H. Mori, Notes on the stability of minimal submanifolds of Riemannian manifolds, Yokohama Math. J. 25 (1977), 9–15.

[Mori2] H. Mori, A note on the stability of minimal surfaces in the 3–dimensional unit sphere, Indiana Univ. Math. J. 26 (1977), 977–980.

[Mu1] Y. Muto, The Gauss map of a submanifold in a Euclidean space, J. Math. Soc. Japan 30 (1978), 85–100.

[Mu2] Y. Muto, Isometric minimal immersions of spheres into spheres isotropic up to some order, Yokohama Math. J. 32 (1984), 159–180.

[Mu3] Y. Muto, Some properties of spherical harmonics, Yokohama Math. J. 32 (1984), 181–189.

[N] R. Naka, Some results on minimal surfaces with the Ricci condition, pp. 121–142 in *Minimal Submanifolds and Geodesics*, M. Obata, ed., Elsevier/North–Holland, New York, 1979.

[Ni1] J.C.C. Nitsche, Elementary proof of Bernstein's theorem on minimal surfaces, Ann. of Math., (2) 66 (1957), 543–544.

[Ni2] J.C.C. Nitsche, Some new results in the theory of minimal surfaces, Bull. Amer. Math. Soc. 71 (1965), 195–270.

[Ni3] J.C.C. Nitsche, A new uniqueness theorems for minimal surfaces, Arch. Rat. Mech. Anal. 52 (1973), 319–329.

[O1] R. Osserman, Proof of a conjecture of Nirenberg, Comm. Pure Appl. Math. 12 (1959), 229–232.

[O2] R. Osserman, On the Gauss curvature of minimal surfaces, Trans. Amer. Math. Soc., 96 (1960), 115–128.

[O3] R. Osserman, Minimal surfaces in the large, Comment. Math. Helv. 35 (1961), 65–76.

[O3] R. Osserman, On complete minimal surfaces, Arch. Rat. Mech. Anal. 13 (1963), 392–404.

[O4] R. Osserman, Global properties of minimal surfaces in E^3 and E^n, Ann. of Math. (2) 80 (1964) 340–364.

[O5] R. Osserman, Global properties of classical minimal surfaces, Duke Math. J., 32 (1965), 565–573.

[O6] R. Osserman, *A Survey of Minimal Surfaces*, Van Nostrand Reinhold, New York, 1969.

[O7] R. Osserman, Minimal varieties, Bull. Amer. Math. Soc. 75 (1969), 1092–1120.

[O8] R. Osserman, A proof of the regularity everywhere of the classical solution of Plateau's problem, Ann. of Math. 91 (1970).

[O9] R. Osserman, Minimal surfaces, Gauss maps, total curvature, eigenvalue estimates, and stability, pp. 199–228. in *The Chern Symposium 1979*, W.–Y. Hsiang et al ed., Springer–Verlag, New York, 1980.

[P–T] C.–K. Peng, and C.–L. Terng, Minimal hypersurfaces of spheres with constant scalar curvature, *Seminar on Minimal Submanifolds*, E. Bombieri ed., Princeton Univ. Press, Princeton, 1983, 177–198.

[P] M. Pinl, Uber einen Satz von G. Ricci–Curbastro und die Gaussche Krummung der Minimalflachen I, II, Arch. Math. 4 (1953), 369–373; 15 (1964), 232–240.

[P–Z] M. Pinl, and W. Ziller, Minimal hypersurfaces in spaces of constant curvature, J. Diff. Geom. 11 (1976), 335–343.

[Ra] J. Ramanathan, Harmonic maps from S^2 to G(2,4), J. Diff. Geom. 19 (1984), 207–219.

[Raw] J. Rawnsley, Twistor methods, pp. 97–133 in *Differential Geometry – Proceedings of the Nordic Summer School*, Lecture Notes in Math. Vol. 1263, Springer–Verlag, 1987.

[R] E.R. Reifenberg, Solution of the Plateau problem for m–dimensional surfaces of varying topological type, Acta Math. 104 (1960), 1–92.

[R–T] H. Rosenberg and E. Toubiana, Complete minimal surfaces and minimal herissons, J. Diff. Geom. 28 (1988), 115–132.

[S–U] J. Sacks and K. Uhlenbeck, The existence of minimal immersions of 2–spheres, Ann. of Math. 113 (1981) 1–24.

[S1] S. Salamon, Topics in four dimensional Riemannian geometry, pp. 33–124 in *Geometry Seminar "Luigi Bianchi"*, Lecture Notes in Math. Vol. 1022, Springer–Verlag, 1983.

[S2] S. Salamon, Harmonic and holomorphic maps, in *Geometry Seminar "Luigi Bianchi"*, Lecture Notes in Math. Vol. 1164, Springer–Verlag, 1985.

[S] R. Schoen, Uniqueness, symmetry, and embeddedness of minimal surfaces, J. Diff. Geom. 18 (1983), 791–809.

[S–S–Y] R. Schoen, L. Simon and S.T. Yau, Curvature estimates for minimal hypersurfaces, Acta Math. 134 (1975) 275–288.

[Si] J. Simons, Minimal varieties in Riemannian manifolds, Ann. of Math. 88 (1968), 62–105.

[Sp] J. Spruck, Gauss curvature estimates for surfaces of constant mean curvature, Comm. Pure Appl. Math. 27 (1974), 547–557.

[T] T. Takahashi, Minimal immersions of Riemannian manifolds, J. Math. Soc. Japan 18 (1966), 380–385.

[U] K. Uhlenbeck, Harmonic maps into Lie groups (classical solutions of the Chiral model), preprint.

[Va] G. Valli, On the energy spectrum of harmonic two–spheres in unitary groups, Topology 27 (1988), 129–136.

[V] K. Voss, Uber vollstandige minimalflachen, L'Enseignement Math., 10 (1964), 316–317.

[W] N. Wallach, Minimal immersions of symmetric spaces into spheres pp. 1–39 in *Symmetric Spaces*, Short Courses Presented at Washington University, W.M. Boothby and G.L. Weiss ed., Marcel Dekker, New York, 1972.

[We] H. Wente, Counterexample to a conjecture of H. Hopf, to appear, Pac. J. Math.

[Wo1] J. Wolfson, On minimal surfaces in a Kähler manifold of constant holomorphic sectional curvature, Trans. of Amer. Math. Soc. 290 (1985), 627–646.

[Wo2] J. Wolfson, Harmonic sequences and harmonic maps of surfaces into complex Grassmann manifolds, J. Diff. Geom. 27 (1988), 161–178.

[Wu] H. Wu, The Bochner technique, Proc. of 1980 Beijing Symp. Vol. II, S.S. Chern and Wu Wen–tsun, eds., Science Press, Beijing 1982.

[X1] F. Xavier, The Gauss map of a complete non–flat minimal surface cannot omit 7 points on the sphere, Ann. of Math. 113 (1981), 211–214.

[X2] F. Xavier, Erratum to previous paper, Ann. of Math. 115 (1982), 667.

[Y1] K. Yang, Homogeneous minimal surfaces in Euclidean spheres with flat normal connections, Proc. Amer. Math. Soc. 94 (1985), 119–122.

[Y2] K. Yang, Frenet formulae for holomorphic curves in the two quadric, Bull. Austral. Math. Soc. 33 (1986), 195–206.

[Y3] K. Yang, *Almost Complex Homogeneous Spaces and Their Sub–manifolds*, World Scientific, Singapore–New Jersey, 1987.

[Y4] K. Yang, Horizontal holomorphic curves in Sp(n)–flag manifolds, Proceed. Amer. Math. Soc. 103 (1988), 265–273.

[Y5] K. Yang, *Compact Riemann Surfaces and Algebraic Curves*, Series in Pure Mathematics, Vol. 10, World Scientific, Singapore–New Jersey, 1988.

[Y6] K. Yang, Meromorphic functions on a compact Riemann surface and associated complete minimal surfaces, to appear, Proceed. Amer. Math. Soc.

[Ya1] S.T. Yau, Harmonic functions on complete Riemannian manifolds, Comm. Pure and Applied Math. 28 (1975), 201–228.

[Ya2] S.T. Yau, Some function theoretic properties of complete Riemannian manifolds and their applications to geometry, Indiana Math. J. 25 (1976), 659–670.

[Ya3] S.T. Yau, A general Schwarz lemma for Hermitian manifolds, Amer. J. Math. 100 (1978), 197–203.

INDEX

Adapted frame along f, 9, 11, 96
Ad–invariant inner product, 51
Algebraic Gauss map, 3, 23
Almgren, F.J., 1
Almost complex structures on the twistor bundle, 144
Analytic type, 56, 58
Antiholomorphic, 102
Anti–self–dual, 150
Associated holomorphic curve, 17, 91
Associated minimal surface, 103, 104, 110, 131, 132
Atiyah, M., 149

Bernstein–Osserman theorem, 41
Branched minimal surface, 81
Branching number, 42
Bryant, R.L., 46, 68, 69, 75
Bundle of orthonormal frames, 48

Calabi, E., 23
Canonical divisor on a compact Riemann surface, 30, 35
Canonical form, 141
Cartan–Killing form, 120
Cartan's lemma, 11, 57
Catenoid, 31, 34
Cauchy–Riemann equations, 34
Chen, C.C., 32
Chern–Osserman theorem, 3, 25, 28
Chow's theorem, 23
Classification of minimal two–spheres, 59, 60
Clifford torus, 78
Compatible Riemannian metric, 4
Complete metric, 26
Complete Riemannian manifold, 25
Conformally equivalent, 5
Completely integrable left–invariant distribution, 113, 115
Complex connection forms, 84
Complex flag manifold, 89
Complex torus, 33, 107
Compact Riemann surface, 30, 35, 93
Conformal class, 5
Conformally equivalent, 5
Conformal minimal immersion, 9
    from a compact Riemann surface, 13
Conformal structure, 4

Conjugate minimal surface, 16
Connection, 5
Costa's surface, 33
Covariant derivative, 8
$\mathbb{C}P^n$, 83

Darboux coframe, 52, 98
Darboux frame, 52
De Giorgi, E., 1
Degree of a holomorphic curve, 24, 93
Din, A.M., 5
Directrix curve, 46, 60, 81
Douglas, J., 1
Dual frame, 7

Eells, J., 80
Eells–Salamon theorem, 148
Elementary argument principle, 63, 116
Embedded minimal surface, 31
Energy functional, 104
Enneper's surface, 32
End of a complete minimal surface, 23, 30
Equidistribution property, 24
Euclidean connection, 9
Euclidean frame, 10
Euler–Poincare characteristic, 63, 116
Extended vector field, 10
Exterior bundle, 7
Exterior derivative, 6
Extrinsic invariant, 114

Federer, H., 1
Finitely connected surface, 28
First variational formula, 2
Fleming, W.H., 1
Frenet bundle, 59, 104, 107
    over a minimal two–sphere, 59
    over a holomorphic curve, 89, 91
Fubini–Study metric, 20, 83
Fujimoto's theorem, 44
Fundamental theorem of Riemannian geometry, 5, 6

Gauge field, 151
Gauss–Bonnet–Chern theorem, 63, 86
Gaussian curvature, 6, 7, 12, 84
Gauss map of a conformal immersion, 17

Generalized conformal minimal immersion, 103
Generalized minimal surface, 81
G–flag manifold, 110
G–invariant complex structure on G/T, 124
Globally defined symmetric form, 54, 146
GL(n;ℝ)–principal bundle, 47
Grassman manifold of oriented two–planes, 17
Green's function, 25, 27
$G_2(TN)$, 148
𝔤–valued left–invariant 1–form, 49

Harmonic map, 104
Harmonic sequence, 109
Helicoid, 32
Hermitian metric, 4
Hodge star operator, 7
Hoffman–Meeks theorem, 33
Holomorphic 1–form, 15
Holomorphic invariants of a minimal two–sphere, 57, 58
Holomorphic line bundle, 120
Holomorphic Gauss map, 2, 14, 17
Holomorphic symmetric differential, 107, 118
Holomorphic sectional curvature, 21, 83
Homogeneous embedding, 115
Horizontal bundle over G/T, 124, 130, 135
Horizontal curve, 67, 75, 125
    in SO(m)–flag manifolds, 122
    in Sp(n)–flag manifolds, 134
Horizontal distribution, 66, 110, 124
    on the twistor bundle, 143
Horizontal lifting, 147
$\mathbb{H}P^1$, 73
$\mathbb{H}P^n$, 65, 132
Huber's theorem, 28
Hyperbolic Riemann surface, 25
Hyperelliptic Riemann surface, 40
Hyperquadric, 17

Immersion theorem for punctured compact Riemann surfaces, 35, 36
Induced complex structure, 4
Induced singular metric, 121
Invariant almost complex structures on SO(2n)/U(n), 139
Invariant metrics, 51
    on G/T, 124, 130, 134
    on SO(2n)/U(n), 140, 143
Instantons, 149, 151
Invariant symmetric differential, 100, 101, 118
Isolated zeros, 56
Isometric deformations of a minimal surface, 23
Isothermal coordinates, 2, 4

Isotropic surfaces, 146
Isotropy representation, 47
    of U(1)×U(n), 87
Isotropy subgroup, 47, 48

Jordan curve, 28
Jorge–Meeks theorem, 31

Kahler form, 82, 84, 94
Korn–Lichtenstein theorem, 3

Lawson's minimal surface, 78, 79
Laplace–Beltrami operator, 2, 7, 25, 42, 52, 92
L(D), 35
Levi–Civita connection, 5, ,6, 51, 54
Lie algebra, 47
Linear frame bundle, 47
Local normal form for a holomorphic curve in Q_2, 114

Maurer–Cartan form of SO(n), 18
    of U(n), 21
    of Sp(n), 71
Maurer–Cartan structure equations, 50
Maximal principle for harmonic functions, 3
    for subharmonic functions, 63, 93, 116
Maximal torus, 110
Mean curvature vector, 2, 11, 53
Meromorphic 1–form, 30
Metric connection, 10
Metrics on the twistor bundle, 142, 143
Metric structure equations for a holomorphic curve, 92
Minimal cone, 46, 76
Minimal surface, 1, 13
    of infinite total curvature, 33
    of arbitrarily large degree, 35
    two–sphere, 52
    in a Kahler manifold, 95
Moduli of minimal $\mathbb{C}P^1$, 80, 108
Moving frames, 47, 50

Nondegenerate holomorphic curve, 86
Nondegenerate horizontal curve, 125, 130, 135
Non–parametric minimal surface, 41
Normal bundle, 61
Normal curvature forms, 11, 54
Normal Gauss map, 20
Number of punctures, 39

$\mathfrak{o}(n)$–valued Maurer–Cartan form, 10
$O(\mathbb{R}^n)$, 48
Orientation–preserving orthogonal complex structure, 137, 140
Orientation–reversing orthogonal complex structure, 151
Orthogonal Plucker formulae for SO(2n), 128
 for SO(2n+1), 130
Orthogonal twistor bundle, 137, 141
Orthonormal coframe, 4, 5, 6
Osculating metric, 93, 121
Osserman's theorem, 41

Parabolic Riemann surface, 25
Penrose fibration, 46, 64
Penrose, R., 149
Period condition, 15
Picard's theorem, 26
Plucker embedding, 90
Plateau problem, 1
Plucker formulae, 94
Polar divisor of a meromorphic function, 39
Poles, 30, 31
Positive roots, 124, 129, 133
Principal divisor, 37
Pseudoholomorphic immersion, 61
Punctured compact Riemann surface, 3, 23

Q_n, 17, 110
Quantization theorem, 92
Quaternionic conjugation, 71
Quaternionic flag manifold, 133
Quaternions, 64, 132

Rado, T., 1
Rational normal curve, 78, 107
Regular point, 89
Reifenberg, E.R., 1
Residue, 37
Ricci equations, 12
Riemann–Hurwitz relation, 30, 42
Riemann–Roch theorem, 35, 40, 55
Rigid motion group, 48
Root spaces of SO(4), 120
 of SO(2n), 124, 138
 of SO(2n+1), 128
 of Sp(n), 133

Scherk's surface, 33
Second fundamental forms of an immrsion, 11, 53
Self–duality, 150, 151
Shoen, R., 3
Simple roots, 124, 129, 133
Simply connected minimal surface, 16, 19
Singular divisor, 85, 94, 121
Singular Hermitian metric, 85, 93, 121
Smooth analytic type function, 56, 58, 88, 101
Smooth variation, 1
SO(n)/SO(2)×SO(n–2), 18
SO(n;$\mathbb{R}$), 48
SO(S^n), 49
SO(2n)/T, 110
SO(4)/SO(2)×SO(2), 111
SO(2m+1)/SO(2)^m, 60, 128
SO(2n)/SO(2)^n, 123
SO(N)/U(n), 137
SO(2n)/U(n), 137
SO(4)/U(2), 140
SO(6)/U(3), 140
SO(3)/SO(2), 119
Sp(n), 65, 132
Spin(n), 65
Sp(n+1)/Sp(1)×Sp(n), 65, 132
Sp(n)/Sp(p)×Sp(n–p), 132
Sp(n)/U(n), 132
Space of constant holomorphic sectional curvature, 82
Stable minimal immersion, 1
Stereographic projection, 20
Structure equations for minimal $\mathbb{C}P^1$ in $\mathbb{C}P^n$, 103
SU(2)/U(1), 140
SU(4)/S(U(1)×U(3)), 140
Subharmonic function, 25
Superhorizontal distribution, 110
Superminimal surfaces in Q_2, 118, 119
Symplectic frame, 74
Symplectic Plucker formulae, 136

Total curvature, 22, 29
Totally geodesic, 56
Total ramification index of a holomorphic curve, 93
Torsion forms, 82
Twistor fibration, 64, 66
Twistor lift, 148
Type (1,0) connection, 82
Type I inner symmetric space, 110

U(1)×U(n)–principal bundle, 86
U(1)^2×U(n–1)–action, 88

U(1)×Sp(n)–bundle, 66
U(2)/SO(2)^2, 119
U(n)/U(1)×U(n–1), 20, 21, 83, 99
U(n), 20, 21, 82
Umbilic point, 56
Underlying Riemannian metric of a Kahler manifold, 96, 97
Unitary coframe, 82, 83

Veronese surface, 77
Voss' surface, 45

Weingarten equations, 10
Weierstrass ℘–function, 33, 107
Weierstrass point, 40
Weierstrass representation, 19, 20
Weyl tensor, 150
Wirtinger;s theorem, 24
Wood, J.C., 80

Xavier's theorem, 44

Yang, K., 36, 116, 136
Yang–Mills functional, 151
Yang–Mills theory, 149

Zakrzewski, W.J., 80
Zero logarithmic capacity, 44